Field Guide for Construction Management

Management by Walking Around

Dennis H. Sanders

iUniverse, Inc.
Bloomington

Field Guide for Construction Management
Management by Walking Around

Copyright © 2011 by Dennis H. Sanders.

The views expressed in this work are solely those of the author and do not necessarily reflect the views of the publisher, and the publisher hereby disclaims any responsibility for them.

iUniverse books may be ordered through booksellers or by contacting:

iUniverse
1663 Liberty Drive
Bloomington, IN 47403
www.iuniverse.com
1-800-Authors (1-800-288-4677)

Because of the dynamic nature of the Internet, any web addresses or links contained in this book may have changed since publication and may no longer be valid. The views expressed in this work are solely those of the author and do not necessarily reflect the views of the publisher, and the publisher hereby disclaims any responsibility for them.

Any people depicted in stock imagery provided by Thinkstock are models, and such images are being used for illustrative purposes only.

Certain stock imagery © Thinkstock.

ISBN: 978-1-4620-6712-1 (sc)
ISBN: 978-1-4620-6713-8 (hc)
ISBN: 978-1-4620-6714-5 (e)

Printed in the United States of America

Library of Congress Control Number: 2011961461

iUniverse rev. date: 12/08/11

Dedication

I would like to dedicate this book to my loving wife Colleen. She has been a work widow and scout widow all of these years, without her support this or any other project that I have been involved in would not have happened.

My Thanks to my colleagues and Peer Reviewers:

Tom Jacobs PE Senior PM SAIC
David Conover CCM CPC VP HDR
Mike Hafling VP CAS Construction

Contents

Dedication   iii

Preface   vii

Chapter 1. Purpose   **1**

Chapter 2. Design Phase Services   **3**

Developing the Construction Documents   3

Constructability/Biddability Reviews   8

Contractor Prequalification   11

Chapter 3. Bid Phase Services   **15**

Pre-Bid Meeting   15

Bid Opening   17

Chapter 4. Construction Phase Services   **21**

Pre-Construction Conference   21

Project Schedules and Reviews   24

Contractor Payment Requests   40

Meetings   42

Managing Resources   54

Documentation   56

Communications   57

Monitoring of Local Landowner Agreements   59

Monitoring Project-Acquired Permits   60

Reports   61

Monitoring of Construction Documents   66

	Training of Owner's Staff	76
Chapter 5.	**Contract Changes**	**79**
	Request for Pricing (RFP)	80
	Field Changes	84
	Change Orders	88
Chapter 6.	**Claims**	**95**
	Delay Claims	95
	Defending against Claims	108
	Expectation of Proof	110
	Documentation of Claims	111
	Claim Prevention	113
Chapter 7.	**Dispute Resolution**	**117**
Chapter 8.	**Testing, Startup, and Commissioning**	**121**
	Pre-Operational/Functional Testing	122
	System Startup and Integration	127
	Commissioning	128
Chapter 9.	**Project Completion and Closeout**	**131**
In Conclusion		**139**

Preface

Success for a construction manger is turning the completed project over to the owner for beneficial use

- In Budget,

- On Time,

- Constructed to the plans and specifications, and

- Properly documented.

The title of this book indicates you can find field applications and solutions for field problems. However to make this work you must tour the site with the contractor's project superintendent on a regular basis. Your best success cannot be had by sitting in a chair, reading inspection reports and waiting for the contractor to tell you what is happening.

Most of you who read this will be schooled in what needs to be done. The aim of this book is to give you ideas for how to accomplish what you are supposed to do. For example how do you know what to do to reclaim a slipped schedule? What are your options? Can you write language into a contract to help solve these kinds of issues?

This book includes many of the author's opinions and methods, all of which have been developed and used successfully on projects. Some of these methods may not be allowed by statute or regulation in some states, counties, or municipalities. You will have to research the local laws and regulations. However, by and large these methods are allowed, and they work. My goal is to share with you what I have learned in forty years as a construction superintendent and construction manager in many states for private and government clients at the project and program levels.

    ** The use of "he" is intended to be generic in nature using he/ she or he or she became so encumbering that I decided to use the generic term he and include this note. There are many fine women in this business, some of whom I have encouraged to go forward. Not because they are women but because they are good at what they do.

Chapter 1

Purpose

Construction management (CM) can be defined very simply as the art and process of guiding a construction project from the design phase through the construction phase to a finished product. The construction manager's responsibility of guiding a construction project to a successful conclusion for the owner is actually quite simple. It consists of four tasks:

1. **Cost:** To see that the project is completed without overspending the budgeted amount (the bid amount plus or minus the dollars added or subtracted in change orders).
2. **Time:** To see that the contractor completes the project per the prescribed contract timetable (the schedule). (The schedule is the contract document time plus or minus time added or subtracted in change orders.)
3. **Scope and Quality:** To oversee the project construction and make certain that the contractor builds the project in conformance to the project documents (the plans and specifications, including modifications made in change orders).
4. **Documentation:** The three tasks listed above may change, and the project documents are modified to reflect the following: (1) changed conditions, (2) the owner's needs, and (3) elements improperly designed or

missed in the design phase. Tracking and documenting progress and these changes in work description and/or scope is as important as the first three tasks.

Cost, schedule, and/or scope may change during the course of the construction phase. Change orders executed by the construction manager as required by design changes to the project documents may affect any or all of the original baseline values that were in place at the beginning of construction. An owner directive may add scope and/or change the intended design resulting in amended plans and/or specifications. That is why a key task of the construction manager is to track the changes in scope, dollars, and time and see that an accurate accounting is made so that at any time during the construction phase everyone knows exactly what the cost, completion date, and current scope are.

The construction manager accomplishes this task by tracking the project documents and the work being completed. All proposals for cost change are documented. Cost change proposals that are accepted and bound into change orders are logged. Changes that require time added to or subtracted from the schedule are documented so that all parties will know how long the contractor has to complete the project. Changes to the design are documented so that all parties know exactly what is to be constructed. Along with the changes, there are submittals, requests for information, correspondence, meetings, and test results that also must be documented, tracked, and filed for reference.

This book will discuss these issues and other responsibilities of the construction manager and the construction management staff. In the following pages are concrete methods for doing these things with the whys and hows to get the job constructed in budget, on time, and to the quality designed in the plans and specifications.
MONITOR:

1. Cost
2. Schedule
3. Scope and quality
4. Documentation

Chapter 2

Design Phase Services

Developing the Construction Documents

Purpose: Understanding the Use of the Contract Documents

The parts of the contract documents held separately are nonentities. When assembled together, they become an active guide to the owner, designer, and contractor. The construction manager is the keeper and enforcer of the contract documents.

The Invitation to Bid

This section of the construction document defines the scope of work, lays out the pricing apparatus for the construction phase of the project, and directs the contractor through the bid process.

The bid bond required with the bid, the bid date and time, the location to turn in bids, and the bid-opening location are all given in this document. The scope of work is defined in this section in terms of bid items that contractor will provide pricing to accomplish. The bid can be requested in several forms, such as:

- **Lump sum**: The entire project is priced in one number.

- **Unit price**: The various components of a project are priced per unit—cy, lf, ea, lb, ton, ton mile, and so on. Usually an estimate of the total quantity is given and blanks are provided for the unit price and an extended total price. The extended price is calculated by multiplying the unit price times the estimated quantity. A total price is obtained by adding all of the extended prices to a total.

- **Cost plus**: Unit costs are provided for labor and equipment, with a maximum markup for overhead and profit added to establish a unit price for each category. Materials are provided at cost plus a standard markup.

- **Cost plus a fixed fee**: Costs are paid at base rates, including overhead, and a fixed percent is added.

There are other delivery methods for estimates of cost. Most bids are delivered lump sum or unit price.

It is important to go over this section of the bid document in the pre-bid conference. This section is where contractor will acknowledge having received and reviewed all addenda and his due diligences to gather information for bid preparation.

The Agreement

This section is the actual terms and conditions of the contract wherein the contractor agrees to enter into a contract and signs the agreement with the owner to accomplish what is defined in all of the other sections of the construction contract document. This section should contain a defined, detailed scope of work, critical dates, and the accurate dollar amounts for which the contractor intends to accomplish the work.

In other words, it is the scope, schedule, and cost for the project. This section binds all of the sections together so all of the construction documents are considered complementary. Signature pages for the owner and contractor are also found in the agreement.

The General Specifications

These are the construction manager's tools. Just as a carpenter uses a hammer to drive nails, the CM uses the general and supplemental conditions to establish and enforce the requirements of the construction phase of the project.

The purpose of the general and supplemental sections of the project documents is to establish the rules under which the stakeholders of the project will operate together to accomplish the common goal of constructing the project that the owner directed the designer to create on paper. The project documents are the guide and plan to deliver a constructed project for a fair price. The general conditions (division 00) of the project document give general and owner-specific policy guidance regarding administrative matters pertaining to the construction of the project.

The supplemental conditions section of division 00 reflects specific policy guidance that modifies the general conditions to reflect specific needs of a specific project. Division 00 also contains the bid documents for the project, including scope of work, bid date and time, insurance requirements, and bond requirements. It also includes the bid form and agreement forms required to be submitted with the bid. General requirements (division 01) may also contain general performance specifications that relate to the project performance overall. The technical specifications (divisions 02–48) define the hard goods, such as concrete, grading, structural products, machinery, and equipment and give specific direction as to the construction and delivery of those goods.

The general and supplemental sections of the specifications are the tools of the construction manager so he or she may be asked to write or assist in writing all or some sections of either or both of them. Always review the sections pertaining to changed field conditions, terminations, order of precedence, submittals, claims, owner's responsibilities, contractor's responsibilities, and owner representative's responsibilities, payment, contract change, retainage, bonds and insurances, safety, and project construction schedule. The construction manager must understand how the bid and construction contract format were established. The construction manager and staff must know these sections very well, better than anyone else on the project. The construction management staff will rely on these sections of the contract document as the basis for administering the construction contract. The construction management staff will "work for" the project document. That is to say the construction contract documents must guide the construction manager in all decision making because they are the binding contract language.

There will be times when the designer, owner, or contractor may not agree with a request, a change order, a piece of equipment selection, or some other action taken by one party or the other. All decisions and all direction must match with the document or one party or the other will have cause to take legal action. It is the construction manager's responsibility to see that there is no cause for such action by reminding the parties of the language and direction given in the documents.

While reviewing or writing any section of the general and/or supplemental specifications, the reviewer/writer must understand that federal and state law and county and municipal ordinances govern and are the bases for much of the language in the document. General construction practice and current technology will also be considered when writing the specifications. Tradition, standard practice, local custom, and accepted practice may be considered but must not be the guiding principles of the document. If law, ordinance, regulation, technology, or standard practice change, then the document must be reviewed and possibly modified to reflect the change(s). However, the project document requirements may be more stringent than the guiding documents. Then, except in cases that are unfair to the contractor, the project document will be precedent to law or ordinance.

The Supplemental Specifications Modify the General Specifications

The general specifications are usually a standard set of "boilerplate" documents that have been produced to guide all sorts of projects, perhaps even vendor purchases. The general specifications may have been produced by a governmental agency, such as the federal, state, or municipal government for which you may be contracted. They may be a standard written by a guiding organization such as AIA or EJCDC. The supplemental specifications make changes to the general specifications to address specific needs, concerns, and issues that are pertinent to the specific project for which they are written. A good example would be the following.

If the standard federal military general specifications were your governing specifications for federal acquisition regulations (FAR), they do not address specific water/wastewater regulations that may vary from location to location, so the supplemental specifications would have to address those specific issues. Locally written general specifications do not address requirements when federal funds are included in projects, so such things as the Davis Bacon Act must be spelled out in the supplemental specifications.

The construction manager may find him/herself dealing with strict schedule considerations, payment requirements, regulations, or needs regarding working times, noise, light, work periods, and so on. All of these and any other specific issues that may or may not be addressed in the general specifications must be addressed in the supplemental specifications.

When developing the general and supplemental specifications:

- Understand that the supplemental specification language modifies the general specification.

- Know the state and local laws and regulations that govern.

- Make the language clear and unambiguous.

- Use strong, fair language to guide the path should issues develop.

- Know that the contract documents govern all project decisions.

Constructability/ Biddability Reviews

The construction manager should have the most knowledge regarding construction methods and practices outside of the contractor's house. The goal of these reviews is to help ensure:

- Consistency

- Completeness

- Clarity

in the project bid documents. It makes the construction manager the most likely member of the project team to review the plans and specifications for problems that may arise during the construction phase. This is most especially true if the construction manager has experience in the construction side of the business. The contractor views most aspects of construction differently than designers or owners do. Generally, contractors have experience with several different designers and must have a broader understanding of the industry as a whole. Contractors do not conceive or interpret; they look at the design they are given in pieces then put the pieces together to develop a method, schedule, and cost. Therefore, the construction manager should be invited to review the construction documents for constructability and biddability issues. There may be more than one review. Very often a review and comment is requested at the 30 percent and/or 70 percent design stages and then a final review and comment at the 90 percent stage.

Such issues as material availability, product suitability, owner direction, soil conditions, equipment accessibility, maintenance access, site storage and lay-down area, scheduling, community impact, labor availability, cost reduction opportunities, and completeness of the design should be included in each constructability review. Subsequent reviews should also include checking to see that previous review comments were addressed properly and adequately. On projects regarded as time critical, make sure the specified scope can be completed within the desired time frame by preparing a baseline construction schedule. Often space constraints will allow only small crews or a limited number of crews to work in a project area. This condition may be cause for concern if the schedule requires large numbers of man hours to be spent in a short duration. The reviewer needs to be careful to only base the review on general construction practices, not on specific knowledge of unique means and methods, unless asked to do so. Specific problem areas of the construction may require certain skills that are not normal or usual. These requirements need to be pointed out to the team so they can be addressed in the process of design.

An example of this might be pile driving. Pile driving with a diesel hammer is noisy and dirty work. There must be ample lay-down area for the piles and time allowed for welding, mobilization, and demobilization. Vibrations caused by the operation may affect nearby structures. If this operation is to be accomplished in a residential neighborhood, it can cause real problems for the folks who live close to the work, so the team will have to do substantial PR work ahead of the pile-driving activity or require alternate methods in the bid document. Upon learning of the lay-down requirements, noise concerns, and vibration concerns, the design team may select a different method of foundation support, such as auger cast piles or caissons.

Look for information in the specifications that give detailed, variable information that is required to build certain aspects of the project. This variable information, such as rebar clearances, can be shown in table format. Other information that might be shown this way is rebar lap, door schedules, paint and/or coating schedules, concrete wall and floor finish schedules, and epoxy usage and depth schedules

Alert the designer if an equipment access will cross a public thoroughfare. The road may not be in the construction site but may be unavoidably impacted and require rehabilitation or total reconstruction because of the weight of the project vehicles and/or activities of the equipment. Be sure that the owner gets what he wants. Often times the roadway owner has had problems and wants to avoid the pitfall again. The issue may have been communicated but not forcefully, so the designer may not have picked it up. It is important the construction manager pays attention to these issues so he can assist the designer to give the owner the best product. By catching such problems that may occur further along in the project may be avoided.

Part of constructability/ biddability reviews is to recognize sequencing issues. Often some pieces of the work need to be completed before other pieces because of constraints that make the work excessively costly to construct in any other sequence. The constraints can be anything from location of subsurface piping runs to weight and size of equipment that requires special lifting considerations. Issues such as these can cost projects substantial amounts of money because expensive specialty equipment, such as special long-reaching or low ground pressure equipment, is required to accomplish the work that would be cut off if it were left until later in the project. This type of information may help the designer to guide the owner in some decision making.

Projected lead times for difficult-to-obtain equipment or materials should be identified to help the designer and owner consider pre-purchasing if the lead times warrant that type of activity. Consider other logistical issues, such as material lay-down and equipment storage on site. These issues can have an adverse affect on the price of a project.

The goal of constructability/biddability reviews is to help prevent construction issues from costing the owner time and/or money.

Can it be

- constructed as designed?

- accomplished without restricting means and methods?

- Understood so that the contractor can provide a meaningful bid on the project as designed?

- Are the bidding documents clear and clean?

Contractor Prequalification

There is no assurance of getting only the best and brightest contractors to bid on any given project. The option of prequalifying contractors may be used to give the owner a high level of confidence that the contractor selected to build his project is capable and knowledgeable. Local or state statutes my prohibit prequalification in some or all forms and/or circumstances. Generally, however, prequalification is acceptable because it is at this level that all contractors have the opportunity to display their capabilities and be judged against each other in a fair and open comparison according to equitable criterion.

Prequalification provides the opportunity for all contractors who desire to construct the project to submit and have their qualifications measured against other submitting contractors and the construction requirements of the project. The prequalification process is open to all contractors that respond to the request for qualification (RFQ). Therefore, it meets all of the free trade requirements. This process also aids the contractors who may not be qualified by allowing them to check their qualifications in a process where cost to submit is minimal compared to the cost of bidding a project. The goal of the prequalification is to restrict the bid submittals to those contractors who are truly qualified to construct the project.

A request for qualification should be prepared that gives a good description of the scope of work to be performed. The RFQ should ask for specific qualifications, such as:

- Contractor's work history, with three-to-five projects of same type, similar scope, and similar size.

- Contractor's organization chart and résumés for the project, with key personnel identified **who would be assigned to the project**. Key personnel would be project manager, project superintendent(s), scheduler, project safety manager, project engineer, and possibly the project environmental coordinator.

- Contractor's claim/litigation history.

- Contractor's local agency license number(s).

- Contractor's financial capability.

- Contractor's safety record, including his experience modification, incident, and severity rates.

- Contractor's experience with controlled insurance programs (CIPs), such as OCIP, PCIP, and/or CCIP, if applicable.

- Contractor's bonding capacity.

- Contractor's insurance coverage limits.

- Contractor's management plan.

- A minimum of three project references, including contacts and phone numbers.

- A maximum number of pages for the RFQ is given. Résumés are limited in size but are not usually charged against the page count.

Minimum criteria should be established in the RFQ for each of these items. Each submitting contractor's qualifications are compared to the criteria, and only the best are allowed to obtain bid documents and compete for the right to be the chosen contractor for the project.

This phase of contractor selection should be advertised in the local newspapers through the owner's process locally, regionally, and if desired nationally in the publications such as *The Dodge Report, AGC* and/or *ABC publications, Engineering News Record,* and so on. Contractors may call to ask questions. Questions are taken in written format only, and addendums to the RFQ are issued answering all questions submitted if warranted. The key to this process is that all answers are given to all respondents to the request for qualifications.

The prequalified short list, while no absolute number is required, should include no fewer than three and no more than eight contractors. Selection may be made via an open or blind process on the basis of best qualified. Generally speaking, in many governmental jurisdictions three is the least number of bidders acceptable to make the bid process competitive. Eight is somewhat arbitrary but from a contractor's point of view is a high optimum number that allows an optimum chance for success. Data suggests that six is the most optimum number of bidders. With more than six bidders, contractors are successful one in ten to twelve times. When bidding projects with six or fewer bidders, the success ratio increases to one in four, but the success ratio depends on the contractor.

References should be used as tie breakers. A blind process would be one in which each contractor's identity is kept confidential by giving each submitter a number, and no identifying documents are passed to the reviewing team. If the selection cannot be made via a blind process, then all participants in the selection process must keep the process totally professional and impersonal. The blind process is used by some State DOTs and by some private owners.

Monitor:

- Key provisions are those by which all proposers will be judged.

- Keep the key provisions fair, clear, and simple.

- The objective is to get the contractors on the bid list that are best suited for the owner's needs and the project requirements.

Chapter 3

Bid Phase Services

Pre-Bid Meeting

Purpose: To meet with the interested or prequalified contractors, tour the project site, discuss the owner's intent for the project and pertinent issues in the bid document, and answer questions that the contractors have regarding the bid documents and/or project.

Process: The interested or selected prequalified contractors are contacted and instructed as to the process for obtaining a set of bid documents. They are also instructed that there will be a pre-bid meeting one to two weeks hence. Attendance at this meeting may or may not be mandatory to be allowed to submit a bid, depending on the desire of the owner and/or designer.

At this point, the project is still in the design phase. Therefore, the pre-bid meeting is conducted by the design manager or his designee. Attendees at the pre-bid meeting physically visit the site, ask questions, obtain answers to questions, and meet the owner, designer, and construction manager.

Contractors may have asked questions via fax, email, or regular mail. All questions are to be directed through the designer and may only be taken from the contractor in writing. Answers to questions will be made through the designer designated to handle all questions. The answers will be given in writing to all contractors. At this point, all contractors present are asked if they have received answers to questions one, two, etc. Sets of these answers should be available in case some have not received the answers yet. All questions from all bidders will be answered to all bidders. Undoubtedly, there will be questions at the pre-bid meeting. Accurate minutes must be kept. All questions asked and answers given at this meeting must be given as part of the meeting minutes to all attendees depending on owner policy and/or the questions asked. The answers may be in an addendum to the bid.

The design manager should include several items on all pre-bid agendas includes:

- Introductions of the owner, design, and construction management staff.

- The bid date.

- Changes to bid date, if any.

- The type of bid and contract—lump sum, unit price, lump sum negotiated, etc.

- Addendums issued to date.

- Anticipated notice to proceed date.

- The duration of the contract.

- Milestones during the construction phase.

- Designer's estimate of cost, if there is one. (optional)

- Bond and insurance requirements.

- Safety requirements; discuss the owner, contractor, partner-controlled insurance program (OCIP, CCIP, PCIP). Answer questions as suits the situation.

- Factors that are important to the owner, designer, construction manager, and contractor.

- Activities, tasks, and/or materials/equipment that may be unusual or unique to this project.

- An attendance sheet.

- A visit to the project site.

- Answers to questions.

If the pre-bid meeting has a mandatory attendance requirement, notify non-attendees of their disqualification.

Bid Opening

In all cases, whether private or public bid openings, the bid opening should be held on the day and date, at the time, and at the location advertised unless a written addendum changes these requirements. Bids will not be accepted after the proscribed date and time. Once bids are submitted and the prescribed time that bids are due has arrived, changes to bids will not be allowed. The specifications need to be very specific that lump sum prices cannot be changed. Lump sum unit price bids will be based on the unit prices unless otherwise directed. If there is an error of extension, then the unit price times the quantity equaling a correct extended price will be used to calculate a new bottom-line price.

In public bid openings, all totals and alternates are read, addendums acknowledged, and bonds and insurances checked. A matrix check sheet form will be provided for each bid opening. The apparent low bidder can be announced with a disclaimer that this is a preliminary finding and that all bids will be checked for responsiveness in pricing and completeness.

All bids are double checked for completeness and responsiveness, at which time a low bidder is established. The successful low bidder is contacted via telephone for a follow-up interview, along with the next two low bidders if allowed and/or necessary. The others are notified that they were not successful. Often public entities do not have the option to take anything other than the lowest bid.

Private entities, however, take the bids, open them, evaluate them, and determine the best course of action. They may inform the apparent low bidder and evaluate other information, such as résumés, schedule, or qualifications to the bid. At that point, the owner may elect to follow the path set below or immediately select the successful bidder.

At this point, the owner may elect to review equipment selected by contractors and compare that list with equipment desired. If the owner, designer, and construction manager are satisfied that the low bidder is responsive in accordance with the bid document, then the low bidder is accepted as the contractor of choice. If there are questions regarding the low bid, the construction manager may, if allowed by regulation and policy, enter into negotiations with the three lowest bidders to establish the best value.

The questions asked during this interview process are innumerable depending on the project scope, where the construction manager and contractor perceive the risks to be, availability of key personnel, named equipment, named subcontractors, etc. At this time, scope may not be changed. From this process, a contractor is selected and notified by issuing a notice to proceed (NTP). The others are sent a letter of appreciation for their participation. Complete notes and attendance records are kept of all negotiation sessions.

If all bids are unresponsive, the owner should reject all bids and if desired, rebid the project. The owner does not have to give a reason for all bids being rejected.

The following apply to all bid openings:

- They are held at the specified time in the specified place, no exceptions.

- Rejected or disqualified bidders are notified.

- If all bids are rejected or disqualified, then all bidders are so notified and a new bid is held.

Chapter 4

Construction Phase Services

Pre-Construction Conference

The pre-construction conference or "pre-con," as it is normally called, is held to establish a relationship between the owner representative, the construction management staff, the designer's point person, and the contractor's management staff. It is also to establish the ground rules for documentation, communications, and protocols, including submittals, requests for information (RFIs), and invoices for the project.

The construction manager schedules and conducts this meeting. The contractor should (but seldom does) present the first draft of his construction schedule, his resource loaded schedule, and a refined draft of his submittal schedule. The following are important items for discussion:

- *project safety*
- project staffing by all parties, including introductions
- documentation format, paper or electronic, and software to be used
- RFI protocols
- submittal protocol
- monthly payment application
- submittal schedule
- schedule of values
- resource loaded schedule
- project construction schedule format
- initial schedule acceptance and payment
- monthly schedule update
- progress meeting schedule
- contract change process, including contingencies and allowances
- mobilization and payment for mobilization
- partial payment schedule and process
- anticipated cash flow projection
- payment for materials on site, stored off site
- cut off day or date for monthly invoices
- permitting

- project photographs

- preconstruction photographs.

- environmental regulations

- land owner rules and stipulations

- testing of soils, concrete, grout, asphalt, welding, etc.

- resident/home owner survey

- traffic control survey and program

- instrument and electrical testing and startup

- mechanical testing and startup

- line of communication for field questions

- the dispute resolution process

As for all formal project meetings, the owner and designer are invited and should have representatives in attendance. In fact, some owners will want to ascertain certain business-level information directly, such as signed bond and insurance forms, list of contractor corporate personnel that will be involved with the project, etc. As all meetings, this meeting will be controlled by agenda, and minutes will be kept and distributed to all attendees. The designer will be there to answer specific design-related questions; the owner will be there to answer questions that relate to his operations that the contractor may have.

If an electronic documentation system is used, see IT staff for electronic process access, software, and training. An IT person should be invited to this meeting.

A few questions about the pre-con meeting that might be worth addressing:

- How long should the meeting last?

- Are minutes of the meeting kept?

- Can all Pre- con participants obtain a copy of the attendance log?

- Who gets the minutes?

- What is owner's role in meeting?

- What is designer's role in meeting?

- Is there a partnering segment of this meeting?

Project Schedules and Reviews

Submittal Schedule

One of the most important documents for all parties participating in the construction phase is the schedule for providing submittal. This information will aid the designer in establishing a firm schedule for his personnel. The designer has more than one project in design for his staff to work on. The better this information is, the quicker the designer can turn around submittals because reviewers can schedule themselves to work on your project. The designer may be requested to provide a list of required submittals. Providing this list to the contractor will aid him in developing the submittal schedule. The construction manager should also review the technical specifications and identify the required submittals to provide a baseline of information with which to compare the contractor's submittal schedule.

The submittal schedule will include all of the required submittals as directed by the project specifications. It will include the submittal's specification section and sequential number for each proposed submittal. It will also include the projected date of submittal. This schedule should anticipate the maximum amount of time allowed by schedule for review. The contractor should also consider what will happen to the schedule if the submittal requires subsequent reviews. Because of the impacts that equipment and materials have on the project construction schedule, it is important that the contractor develop and submit a complete submittal schedule as quickly after the notice to proceed as possible.

The contractor should be required, by specification, to review all submittals before submitting them to the construction manager. It is in the contractor's best interest to submit complete responsive submittals that require only one review. The contractor should check each submittal against the requirements defined in the specifications. If the item submitted is an "equal," he should mark the submittal as such and check the item with even more scrutiny than a named item. Many submittals are returned because the submittal information provided is not complete. Very often had the contractor checked, he would have found the submittal information lacking.

The review of the submittal schedule is to see that-

- all required submittals are included,

- all submittals are submitted in a timely manner as they relate to their required installation time on the construction schedule, and

- the submittal is complete so the designer can schedule the submittal review team to review submittals in a timely manner.

Project Construction Schedule

Purpose: The project construction schedule is the prime tool of the construction manager for assuring that the contractor is in control of the project. The construction schedule is the prime tool of the contractor to see that he is in a profitable mode and in control of his project. Because it is such an important tool, the construction manager may want to tie schedule maintenance to breach of contract. Schedule development and submittal should be incentivized by tying them to payment of early invoices. Such language in the supplemental conditions may be, "Contractor will not receive payment until such time as an accepted schedule is submitted." The construction schedule is the contractor's plan for building the project. Many contractors don't realize that, but it is a fact. If the contractor gives schedule proper consideration during the bidding process and then uses the schedule to guide the project, he will generally lead the project to a successful completion.

Every line item of a schedule is an activity that must be accomplished. Each item is tied to one or more activities that must be completed before it can be accomplished. The succession of all of the activities with durations for a project is called a schedule.

The project document requires that the construction phase of the project begin on a certain date and end on another certain date. That is the schedule duration. Often there are dates within the duration path by which certain portions of the project must be completed. Those dates are milestone dates, and the contractor is contractually obligated to meet them. Those milestone dates are usually tied to liquidated damages clauses to incentivize the contractor to pay strict and close attention to them. There are a certain group of activities that follow each other in succession end to end with no overlap or gap. The group of activities that begins with day one and forms a path through the milestone dates, and end on the last day of the project is called the "critical path" of the project. Any slowing down (slippage) or speeding up (acceleration) of any of the activities along the "critical path" affects the completion date. All other activities are side activities that have some unneeded time before the start date or after the end date before the next connected activity must begin. The unneeded time is called "float."

The construction document may prescribe the schedule format (i.e., hardcopy or digital copy) and/or use of prescribed software, critical path identified, or define other requirements for the project construction schedule. The submitted schedule may not be approved by the construction manager, but it may be rejected if the logic and/or one or more durations along the critical path are obviously flawed. If a review of the project construction schedule demonstrates that there are no obvious flaws and all of the document criteria are met, the schedule will be accepted. The accepted schedule will then be used to measure progress of the project.

Claims for delay cannot be made without an accepted baseline schedule in place to measure progress against. Changes in the critical path caused by the owner can result in an indefensible delay claim. Changes to the critical path may change the entire direction of the project, so monthly reviews are important. The accepted schedule cannot be modified except via change order that allocates more or less time to the project as a whole and probably to a specific activity. Change order time allocations, whether positive or negative, may well affect the critical path.

Delay claims can only be made if the project as a whole is delayed, so delay claims are made against negatively affected critical path activities or side activities that were so negatively impacted that the impact used up the identified float and pushed them on to the critical path. Then it continued to delay the item so that the project duration is prolonged because of the delay.

Each time a change order is issued that affects the duration of any activity or the project as a whole, a revision of the project construction schedule will be produced demonstrating what affect, if any, the change order had on the schedule.

The project construction schedule should end on the end date stipulated in the contract document. If the contract allows for an early completion, and the schedule is accepted, the owner may be penalized for delay if the contractor completes the project after his scheduled end date and it was because somehow the progress was delayed by owner or construction manager issues or activities. Even though contractor reached completion prior to the contract stipulated completion date he may have a claim for delay. It is suggested that the construction document not allow schedules that end earlier than the contractual date in the document.

At the end of the project, the contractor is required to submit an as-built project schedule. The as-built schedule should be a final iteration showing all changes and final effects of delays and/or accelerations on all critical path activities as well as side activities. The as-built schedule is discussed in a later section.

One should always do the following when performing schedule reviews:

- Make sure all contract-required criteria are met.

- Make sure schedule maintenance is tied to breach of contract.

- Check logic for obvious errors.

- Make sure all delayed and advanced items are properly accounted for.

- Make sure all milestones are included and represented on the critical path.

- Make sure the end date of the contract is the end date of the schedule.

Contractor's Short-Term "Look-Ahead Work Schedule"

The contractor is to produce a schedule that demonstrates which schedule activities he will be working on over the next three weeks.

This is called a "three-week look-ahead schedule." This short-term schedule should be detailed, including what days and times critical items of work will be accomplished, including such details as concrete placements, activities that require testing and inspection, activities that require interface with other contractors, and/or the owner's operations. This schedule must be coordinated with the project construction schedule.

How does the three-week look-ahead schedule

- match with the construction schedule,

- coordinate between crews,

- manage holiday periods,

- address milestones, and

- address short critical activity periods (turnarounds)?

Schedule Float

Purpose: To allow both the contractor and the construction manager to have flexibility to manage noncritical path activities (within the realm of their respective responsibilities) as needed.

Schedule float is the difference in time between the early finish date and late finish date of a given schedule activity. The activities on the *critical path* of the schedule have no float or more correctly *zero float.* That is how we know they are critical path items. The float of a given activity belongs to the party that needs it first. Consider the following example.

The contractor is preparing to place some remote slabs on grade. They are scheduled to be completed (early completion) on May 5. It rains for several days, so they are not completed until May 10, but the late completion date is June 15. The contractor uses five days of the forty-one days of float. If the owner delays those same slabs ten days to allow access for a heavy load so the slabs are not completed until May 20, the owner has laid claim to ten of those float days. Because the sum total of the delays to the slabs is less than the forty-one days of float in the schedule, the delays have no affect on the overall schedule for completion. In fact, twenty-six days of float remain for that activity.

Float can be handled in one of three ways:

1. Float belongs to the contractor. After all, he developed the schedule and built in the float for his benefit.
2. Float belongs to the owner. After all, he owns the project, and everything on it is his.
3. Float belongs to both parties and is used as needed to best satisfy the needs of the project. The party that needs the float first takes ownership, with the construction manager being the arbiter. The project is the objective and the first importance, so sharing the float is what is recommended.

Some examples of how float is used follow:

1. Contractor has to build a wall; the wall will take 21 days by duration on the schedule with a late start showing 14 days of float. Contractor submits, receives approval on the block in a timely manner but the supplier has a plant shut down and is 7 days late delivering the block. The activity slips 7 days but will be completed 7 days before the succeeding activity so the contractor gets to use 7 days of the float, no penalty.
2. Contractor has to build a wall, the wall will take 21 days by duration on the schedule with a late start showing14 days of float. The owner changes his mind and wants a different color of block that requires a special run of block that will delay the delivery by 10 days. The activity

slips 10 days which leaves 4 days before the succeeding activity must begin. In this example, the owner gets to use 10 days of the float, no penalty.

However, if these issues are combined for a seventeen-day overall delay, the activity would be pushed into the succeeding activity. If the succeeding activity is a critical path activity, the project would be adversely affected by three days. The offending party would most probably be the last party to claim against the float.

Negative float is the time past the scheduled late completion date that an activity is expected to run before actual completion. That means that the activity will be late being completed and may delay the project. The baseline project schedule will have no negative float. Late items that have delayed the project must be accounted for in a monthly report, and a plan to reclaim the lost time must be submitted. If negative float occurs in a critical path item, a project delay has occurred and requires the construction manager's immediate attention.

Construction schedule:

- Does the construction schedule demonstrate float?

- Do updates demonstrate change in float for items that either finish late or early?

- Does the schedule show any items with negative float?

- Does the contract document address distribution and use of the float time?

Schedule Slippage

If the project construction schedule is the contractor's plan of execution, what happens if the plan goes awry? One of the key roles of the construction manager is to keep that from happening by monitoring the project closely as it progresses forward. One of the first signs of trouble is that the three-week schedule (see "Contractors' Short-Term Look-Ahead Work Schedule") begins to move crews off of the critical path and onto side activities. Daily observation by the inspectors and the construction manager will catch these kinds of adjustments. You may also compare the manpower projections and actual loading to see if contractor is properly manning the job. This is usually because those activities are coming onto the schedule and contractor does not have manpower or equipment or both to execute the schedule. When that happens, the construction manager must act swiftly.

The first action should be to meet in his weekly walk with the superintendent and talk about what he has observed. If that doesn't get action, then a memo may be sent to document the reminder that contractor appears to be slipping on schedule. If slippage continues, a formal meeting may be set up to stir up the enthusiasm of the contractor to maintain the schedule. Should the contractor not respond with action, the contract document must be enforced.

Schedule maintenance should have strong language giving the construction manager the ability to take command if necessary to get the project back on schedule. Some documents allow portions of the work to be removed from the contractor's scope so he or she can concentrate on schedule critical (critical path) activities. Usually the contractor is given some latitude for slippage fourteen to twenty-one days before a formal notice is issued requiring the contractor to present a formal plan of remediation. The more time it takes to get contractor started on a plan, the further he is slipping, so an early attack is essential.

Any and all remediation is at the contractor's cost. Seeing that the project is completed on time is one of the four key responsibilities of the construction manager, so late completion is not an option for success as far as the construction manager is concerned. Allowing the project to get to penalty is an unsuccessful completion of a project.

Review:

- How is schedule slippage tied to breach of contract in the contract document?

- Be conscious of the three-week look-ahead schedule and check it against the construction schedule regularly.

- Walk the site and be in touch with contractor's progress.

- If a contractor begins to slip on schedule, warn him and offer helpful suggestions if you can.

- Know the area you are working in for such things as craft worker availability, equipment availability, good support vendors, and consultants for the sort of work activities the contractor is struggling with.

Schedule Remediation

If the contractor slips on schedule and needs to reacquire the lost time, he or she will prepare a remediation schedule. Some may not feel so inclined and may require some level of coercion to develop the remediation schedule. Be sure that the construction contract has language to obligate contractor to remediate lost time. There are several methods to regain schedule.

Regaining schedule can be a tricky business. Some of the factors that must be considered when determining the best method for remediation are the following:

- jobsite manpower requirements and area available to work,

- availability of trained workers in the area,

- availability of equipment to support an increased work force,

- availability of leadership to manage an increased work force,

- contract mandates for work hours, noise, and light pollution restrictions,

- season and length of workable daylight,

- availability of materials to support an increased work force, and

- effects of longer working hours on the existing workforce, management, and equipment.

Simply adding people may not be viable if there is not room enough to work them effectively. Additional trained workers may not be available, and additional procurement may have to take place, as well as adding more management to lead the added workers.

Working longer hours is usually a serious consideration but can be a dangerous path to follow. Business Roundtable developed information regarding the effectiveness of short-, medium-, and long-term use of overtime. The Business Roundtable publication C-2 "Scheduled Overtime Effect on Construction Projects" demonstrates the loss of efficiency as workers are required to work longer hours. Why does overtime cause loss of efficiency? The human body gets tired, not only the muscular and skeletal portions but also the mental portion. As fatigue sets in, rework becomes more common, and absenteeism grows at a rapid rate. Along with loss of efficiency, the other deterrent is equipment maintenance time becomes scarce, and equipment availability may drop below acceptable levels.

Sometimes a second shift is the best consideration, but that requires, again, more workers and management. The added shift reduces available maintenance time and may not be allowed by the contract without permission. Adding equipment that allows more efficient use of manpower is an option but can require skill levels that may or may not be available in the existing work force.

As the construction manager reviews the contractor's remediation plan, he or she must consider all of these ramifications and discuss them with the contractor. If the contractor-selected method seems risky, it is his risk. Monitor the plan as it is put into motion. If it fails, the construction manager may be looking for other remedies that are in the contract, such as removing work from the contractor's scope. That is heady stuff and has dangers of its own.

There are many ways a contractor can reclaim lost time. Some are listed here:

- Add manpower.

- Add equipment.

- Work more hours (second shift as needed and/or allowed).

- Add work facilitation systems such as gang forms or develop templates to make the work easier to perform.

- Subcontract more of the work

Be careful. As stated above adding more hours is not always a good solution. Workers get fatigued and lose productivity if worked over long periods of time using lengthened working days. Adding more equipment hours and taking away some of the need for manpower is always good. However, if the contractor adds equipment or works equipment more hours, more maintenance must be considered also. Going to around-the-clock work situations does not leave maintenance time, and equipment begins to suffer, so productivity is lost. The construction manager must look at these factors as he reviews the contractor, remediation plan.

Resource Loaded Schedule

Purpose: To explain what resource loaded schedules (RLS) are, how they work, why we require them, and how we use them.

The project construction schedule is the prime tool of the construction manager for assuring that the contractor is in control of the project. Because it is such an important tool, you may tie schedule maintenance to breach of contract in the construction documents. The resource loaded schedule allows the construction manager to monitor contractor's use of resources to maintain the accepted schedule.

The resource loaded schedule is provided with the project schedule at the beginning of the project. It is the contractor's anticipated dollar, manpower, and/or equipment usage over the duration of the project. This schedule works in conjunction with the contractor's daily manpower and equipment report. The resource loading may be incorporated in a packaged scheduling program. These packaged programs integrate the dollar, manpower, and equipment usage to develop current schedule status. The resource information may be a standalone accounting of resource usage.

A certain level of effort must be expended to maintain the scheduled progress toward completion. Contractors as a whole are very good at estimating how many man hours will be required to construct a project. They are also very good at knowing what equipment is needed to execute the construction plan (the schedule) that they have developed. In the RLS, the contractor tells us what he believes those numbers are for each activity or task and over each time period. By monitoring the RLS, the construction manager will know if he substantially increases or substantially decreases his level of commitment.

Usage of a resource, be it money, man hours, or equipment hours, is called the *"burn rate."* If a project is projected to take one hundred thousand man hours and is scheduled to complete in ten months, the "man-hour burn rate" needs to be ten thousand man hours per month average to complete the project on time. An ideal schedule will show the burn rate as a bell-shaped curve. An actual schedule will reflect a curve that looks more like a mesa, with a long, fairly flat period between the ramp up and ramp down periods at each end.

If a contractor lowers his crew size (thus lowering his burn rate) and then begins to slip on schedule, we have the RLS to demonstrate the probable cause for his slippage and can assist him by raising the issue. If contractor presents a claim for delay, the construction manager can go to contractor's manpower usage and compare it to his initial commitment and determine if his change in manpower burn rate was an essential cause for the slippage. This same reasoning is used for a comparative study of project dollars and/or equipment usage. The dollars are usually tracked on an S-curve where the beginning and the end of the project are relatively flat and the balance the dollars track along a sloping line with each month the total amount of dollars consumed are tracked.

Resource loaded schedules can

- help the construction manager track the project progress,

- tell a tale of asset management,

- allow the construction manager to be proactive in determining schedule performance, and

- work as a tool to protect the owner from fraudulent delay claims.

Review of the Resource Loaded Schedule

Purpose: To understand the contractor work plan as it is laid out in the project construction schedule and how that relates to the cost of the individual activities and component equipment defined in the project construction schedule.

Contractor may also be required to submit a resource loaded schedule. This schedule quantifies the dollars to be spent by time period, manpower by craft and quantity, the type of equipment resources, and/or the manpower that the contractor is planning to commit to the project. This document also represents contractor's means and methods. Therefore, it is not subject to public review under the freedom of information act because it is proprietary information.

The construction manager should carefully review the RLS in comparison to the project schedule to see how the manpower and equipment loadings correlate to the ebb and flow of activity on the project. The logics are interdependent. For example, as activities relating to electrical and instrumentation work come online and progress forward, the number of electricians must rise.

The construction contract should request that the contractor report the manpower usage by craft. This report can be requested with the pay estimate each month.

The loading of cost against each item and duration allows the development of a cash flow projection for the project that may serve many purposes, including assisting the owner in utilizing his monies more efficiently.

Tying equipment usage to the construction schedule is an aid to the construction manager as he evaluates the contractor's commitment of resources to the project over the duration. Should the project progress slow and begin to falter, the construction manager can compare equipment on site versus the equipment usage projected in the schedule.

Just as the man hour loading allows the construction manager to monitor the contractor's labor burn rate, monitoring equipment on site aids the construction manager is another method to monitor schedule progress.

The resource loaded schedule

- demonstrates means and methods and are not subject to public review,

- is tied to the construction schedule, and

- provides information regarding the resources to be used on the project.

Schedule of Values

Purpose: To define the process by which the dollar value of each project will be broken into its component parts.

Contractor is required to develop a schedule of values. This is a list of the activities and/or items of material and equipment for which the contractor will seek payment on a periodic basis, usually monthly.

The schedule of values will coincide with the project construction schedule list of activities. The contractor will assign each activity on the project construction schedule a dollar value. The cumulative amount to which this list of dollar amounts equals is the total value of the contract. The contractor is not allowed to front load or unbalance the schedule of values. That is to say contractor is not to create artificially high or low values for any activity or group of activities in the schedule of values. The schedule of values dollars are the dollars in a resource loaded schedule. The schedule of values is necessary when a resource loaded schedule is not required and to serve as a basis of payment.

The project specifications allow payment of a restricted percent of the contractor's contract for mobilization, which should be reflected in the schedule of values.

The contractor should be required to submit the schedule of values for review; the construction manager will send the schedule of values to the project scheduler and estimator for review. If they do not accept it as submitted, they may reject it outright or ask that it be amended according to their comments and resubmitted. When the schedule of values meets all contract conditions and is balanced, it will be accepted and is returned to contractor as acceptable. The contractor may submit only a set number of invoices before the schedule of values is accepted. If contractor has not completed the cost-loaded project construction schedule to a point of acceptability by the time he is through the "grace period" as defined in the contract document to get his schedule of values accepted, he may face working without pay until he gets it accepted.

The schedule of values

- determines pay items and unit rates in a lump-sum project,

- aids the construction manager in managing payment of pay items,

- helps balance payments against schedule progress claims, and

- requires scrutiny to make sure there is no unbalance for upfront payment.

Contractor Payment Requests

Purpose: To define the payment process as it will be accomplished each month.

Contractor uses the approved "schedule of values" to develop a request for payment on a periodic basis, usually monthly.

This list or schedule will coincide with the project construction schedule and is the basis of those periodic progress payments. The responsibility for originating a pay request belongs to the contractor. On or before the predetermined date each month, the contractor's representative will deliver to the construction manager a copy of his updated schedule of values. Each item of the schedule he wishes to be paid for will be marked with a dollar amount and percent of total value. There will also be columns for cumulative dollars to date and cumulative percent paid to date. This form may be generated by the project construction critical path method schedule (CPM) program.

The construction management staff will review the preliminary copy of the payment application as generated by the contractor and set a meeting with the contractor's chosen representative. Those two will go over each item in detail and reach an agreement regarding payment for that payment period. A copy will be made of the agreed quantities, and the contractor will use that as basis for the periodic (monthly) invoice.

A cover sheet form must be completed summarizing the data on the schedule of values into an invoice number including all approved change orders and work completed on them to date, less retainage. The contractor then submits a required number of original invoices with the appropriate signature by the contractor. The invoice has three places for signature: *1. Contractor,* which certifies that he is comfortable with the amount being invoiced. *2. Construction manager,* which certifies that he has checked the invoice for accuracy and recommends the invoice for payment by the owner. *3. Owner,* which acknowledges the invoice and agrees to pay the amount requested in the invoice.

The construction manager's staff will review the invoice for accuracy; if it is accurate, he or she will present it to the construction manager for acceptance. Once signed and recommended by the construction manager, the required number of original copies of the invoice will be sent on to the project controls manager for logging and approval by the owner. The project controls manager will enter the invoice into the owner's payment process. A fully signed and executed copy of the invoice is returned to the contractor for his file and to the construction manager for entry into the tracking log and for the construction manager's site file. Any other copies of the invoice should be distributed throughout the owner's organization as required.

The project construction contract may require that other pertinent information must be submitted to the construction manager before the periodic payment is made. Some of those might be:

- Redline drawing review

- Monthly project construction schedule update

- Project photos

- Other items as required

The contractor payment request should:

- be a product of mutual agreement between contractor staff and construction management staff,

- be processed promptly,

- have enough copies provided so all pertinent parties have a signed original (original signed copies should be provided to the owner, contractor, and construction manager) and

- be accompanied by other items required to be delivered at the time the payment application is delivered.

Meetings

Weekly Progress Meetings

Purpose: To aid in developing the weekly project construction coordination meetings with the contractor, his critical subcontractors, the owner, and the designer. This meeting is held to determine project progress, coordinate activities with the owner and other interested parties, bring issues to the table, document past issue resolution, and maintain project personnel relationships.

The weekly progress meeting is organized managed conducted by the construction manager who has developed an agenda. The agenda should be prepared and distributed at least two days ahead of each meeting and managed by the construction manager or his designee. The construction manager will have present a member of the management staff to keep complete and accurate meeting notes and a sheet listing all of the attendees, who they represent, and other pertinent contact information. This meeting is attended by all of the following:

- the owner's representative,

- the construction manager,

- the designer,

- the contractor,

- relevant subcontractors,

- owner operations staff as they desire,

- others as requested by the construction manager.

This meeting is critical to the ongoing progress and operation of the project. An agenda for this meeting is developed with critical information regarding time completed, time remaining, change orders completed, change orders in process, requests for proposals (RFPs) outstanding, submittal log, and request for information (RFI) log included. Each of these items will be discussed in turn. Also included in the agenda for discussion are issues outstanding from past weekly meetings, new issues, and the three-week look-ahead schedule. This meeting should not last more than an hour.

Complex and/or difficult issues should be dealt with in meetings specific to the relevant issue.

Most of the items on the agenda (see "Weekly Progress Meeting Agenda") will require very little discussion. Issues that arise during the meeting will have an individual assigned to investigate and resolve or report on the issue. The existing issues and other issues, along with the three-week look-ahead schedule, will usually take the most time, as they should.

The three-week look-ahead schedule should particularly include any and all activities that require interface with or disruption of the owner's operations, other contractor's operations, or third party's activities so that coordination can be discussed and preparations can be made to protect the interests of all involved parties.

There are times when vendors are invited because they are in a critical position to impact the progress of the project. Attending parties should be queried for items they want on the agenda. At the meeting, all parties are given copies of the RFI log and the submittal log to review for completeness and accuracy. Items resolved in the last meeting or during the intervening time since the last meeting are to be on the agenda, with resolution noted. No item on the agenda is to be removed unless the resolution to all party's satisfaction has been reached and noted on the agenda.

Complete meeting notes will be prepared and distributed, along with a copy of the meeting attendees. The three-week look-ahead schedule as discussed and agreed upon in the progress meeting will be distributed to all attendees.

The weekly progress meeting should provide all of the following:

- review of critical construction dates,

- construction safety,

- review of near-term construction activity schedule,

- a list all issues that are ongoing for status reports,

- a list of new issues that have arisen since the last progress meeting,

- flow of information, including RFI logs, submittal logs, and contract change logs,

- coordination with other contractor activities, owner activities, activities that interface with the public, critical and/or strategic activities, and vendors, and

- an action item list with assignments for action and required time of completion.

Special Meetings

Purpose: To resolve specific issues that arise during the course of a project.

Special meetings will be held on an as-needed basis. Participants should be requested to submit items for the agenda to the construction manager; the agenda should be distributed at least two days prior to the meeting time and date. These meetings may be open to discuss a topic or issue that may become contentious, to obtain needed information to resolve a dispute, to clear the air and straighten out misinformation and set a path for forward travel, to introduce a new phase of the project, to share information regarding a new work activity, to discuss schedule issues, or to discuss other issues that are pertinent to the project.

Special meetings might be held with a critical vendor that

- is having a problem getting his submittal approved,

- sees future needs of the finished product (plant operation) that the required technology will not accomplish,

- has received never-before-disclosed information that changes the product in a minor or major way, or

- is having scheduling problems, etc. Special meetings are also held to resolve disputes, discuss billing issues, and many other topics.

If a vendor or subcontractor requests a special meeting, the construction manager organizes and controls the meeting. The party requesting the meeting may provide the agenda to the construction manager, but since these meetings are project meetings, they are under control of the construction manager.

All special meetings are handled as follows:

1. They are all controlled by an agenda.
2. Meeting minutes are kept for each meeting. The agenda form is convertible.
3. An attendance sheet is kept for each meeting, even if the meeting is a conference call. Attendees and their companies, phone numbers, and email addresses are noted.
4. Each raised question or issue that requires research is assigned to an individual for research and/or completion. The assignments are written in the meeting minutes.
5. If a follow-up special meeting is required, the time and date is established and noted in the minutes.

After each meeting, the meeting minutes and list of attendees will be distributed to all of the attendees.

Pre-Operation Coordination Meeting

Purpose: To notify all stake-holding parties in an upcoming operation of the nature, duration, and impacts of that operation. It is also to identify and minimize all potential problems at interface points and find common ground for problem solving before problems arise from new operations that involve multiple stakeholders. These types of coordination meetings are particularly important when the project includes or is working in an operating plant and modifications to operating systems or new construction is tying into operating systems. These types of projects require coordinators that work directly with the operations group to schedule down times for tie-ins or equipment or material replacement.

Prior to each activity that will impact the owner or a plant operation; a special meeting should be called with the owner's plant operations supervisor, the plant maintenance supervisor. Any operators and maintenance personnel they feel necessary are invited to attend.

This meeting will be conducted by the construction manager or his designee, an agenda prepared and distributed at least two days prior to the date of the meeting, and attendees invited by the construction manager or his designee. The construction manager will designate someone of the management staff to keep accurate minutes and the attendance sheet for distribution. The contractor or vendor that is to accomplish the tasks that will impact the owner's operation should prepare and present a best schedule of activities that constitute the impacting operation. The group in attendance should then discuss the schedule and reach an agreement on the plan as presented or as amended or agree to further research and discussion based on issues that arise during the meeting. If further discussion in meeting is required, the next meeting date and time should be established prior to meeting adjournment. Each issue is assigned to an individual to research and/or resolve. If resolution is achieved, the individual notifies the construction manager, and the construction manger notifies all of the other parties.

Complete and accurate meeting minutes will be distributed, along with the meeting attendees list and the best proposed work plan and schedule as known at the end of the meeting. Further contact assignments will be noted in the minutes.

Pre-operation meetings are

- held to plan or check plans for a coming critical activity and

- governed by the construction manager and governed by an agenda.

See special meeting checkpoints.

Construction/Operations Coordination Meetings

Purpose: To protect the integrity of the owner's operating system and minimize impacts to the operations of the owner's plant facility during the construction phase of the project.

During the course of every contractor activity that impacts or interfaces with the operation of the plant, coordination meetings will be held on a regular basis, usually daily or perhaps as infrequently as weekly. These meetings, conducted by agenda, are to coordinate the activities of the contractor so that they minimize problems and interruptions that cause disruptions to the owner's plant operations. The construction manager or designee will keep accurate minutes and a record of the attendees that will be distributed to all attendees.

Often during construction the contractor will need some service, such as plant air or plant water, to assist in accomplishing some construction activity. These needs must also be coordinated with plant operations for anything beyond casual, periodic short-term use. Some plant support systems are stressed during normal operations, and a constant draw may diminish the owner's operating capacity.

During times when the plant must take one or more operations offline to accommodate a construction activity, a plant-appointed liaison will be at the activity at all times. This person will be there to facilitate the construction process and to protect the plant systems.

This person will be involved in all discussions and resolutions of problems that arise between the time the shutdown begins and the time the plant is back online at full capacity.

Construction/operation coordination meetings are

- held to plan or check plans for an upcoming critical construction activity that may affect the owner's operations,

- governed by an agenda.

(See special meeting checkpoints.)

Startup Meetings
(Primarily for Plant Facilities)

Purpose: To coordinate and schedule the start-up process for the new facility. All meetings need to be conducted by agenda, with minutes taken.

The initial startup meetings happen early in a project to review the initial startup plan submitted by the contractor. At this stage, the issue is starting the process rolling so someone is thinking about testing, energizing, access, safety, sequencing, and so on during construction so that construction planning prepares for these future events. Looking ahead is a major thought activity that pays huge dividends to the contractor and the project, if it happens. If it doesn't, all too often pain and anguish follow in the form of "we coulda, we shoulda." Why didn't we?

The initial meeting will familiarize everybody with the systems, the testing required, and the process of startup. There will be subsequent meetings to follow up on the development of the startup plan.

When startup is ready to begin, an official startup meeting will be held. This meeting will be called by the construction manager at the request of the lead startup designer. The construction manager will explain the purpose of the startup program and turn the meeting over to the lead startup designer. Attendees are the following:

- lead startup designer,

- startup staff,

- construction manager,

- contractor project superintendent,

- contractor's liaison with the startup staff,

- owner's operation manager,

- owner's maintenance manager,

- owner's area operation supervisor, and

- owner's operation and maintenance people as required

The initial startup meeting will be an overall scope and scheduling meeting. The construction manager will introduce the lead startup designer, and the meeting will be turned over to him for conducting. All subsequent startup meetings will be convened and conducted by the lead startup designer. A master startup schedule will be reviewed. Responsibilities of the startup crew, contractor, inspectors, and owner's personnel as defined in the project documents will be explained and discussed. Generally, the startup crew will require assistance. The assistance is usually one or two electricians, a millwright, a pipefitter, and/or other craft personnel as necessary on call at the convenience of the lead startup designer. They are usually paid for on a time-and-material basis as set up in the construction contract. Meeting notes will be kept and an attendees' sheet, both of which will be distributed to all attendees.

These meetings are usually convened daily at the beginning of the day shift. Problems encountered and their solutions are discussed.

A schedule of the day's activities is handed out, lock-out and tag-out requirements will be assigned, and equipment being started up and tested will be defined. Coordination with the owner's operation is ascertained, and owner interface issues, if any, are resolved. These meetings usually are less than an hour in duration, but they are usually intense.

Construction/operation coordination meetings

- are held to plan or check plans for an upcoming facility startup activity,

- are governed by an agenda,

- pay strict attention to safety standards,

- familiarize all parties with testing and startup requirements and procedures,

- coordinate staff efforts,

- discuss problems and determine solutions, and

- review status and outline daily activities.

Dispute Resolution Meetings

Purpose: To clarify points of difference, give some guidelines for ground rules during dispute resolution, establish lines of communication for discussing points of dispute, and hopefully find a common ground on which to resolve the disputed issues.

The dispute resolution process may already be determined by statute or governing body resolution. If the method has not been prescribed, the owner must determine at an early stage of project planning how he wants to resolve disputes that may arise during the construction phase of his project. There are many options, such as mediation, binding and nonbinding arbitration, or a dispute resolution board.

All of these are methods to resolve disputes and keep them from litigation. A process known as partnering has been developed and successfully used to prevent issues from reaching dispute level. The construction manager should query the owner and be prepared to present the various methods and make recommendations depending upon his experience and level of comfort with each.

If the construction manager prefers, he may bring an experienced party to present the alternatives and explain them to the owner. The ultimate goal in most cases is to keep disputes from litigation. Litigation is costly and can be damaging to the image of the owner.

There are some basic elements that prevail through the processes of the various dispute resolution methods.

Dispute resolution meetings are held at several levels beginning at the project level and ending at the highest level of the owner's governance structure. Most disputes will be solved in the field at the field office. All formal dispute resolution meetings will be controlled by agenda, which will be developed by either party of the point of dispute. Both sides will have opportunity to include items for discussion. Items of dispute should be discussed openly and respectfully. Hostility has no place in a professional setting; often agreeing to disagree is the only solution at a given level in the dispute resolution process. When an impasse is reached at a given level, the issue is handed up to the next level for consideration.

The dispute resolution process begins at the project level. A dispute may arise between an inspector and a foreman or an inspector and a field superintendent. The meeting at this level may occur in the field at the very point of the dispute. The discussion at this level should be civil and based on fact and on contract requirements. If that level cannot resolve the dispute, they refer the issue to their immediate superiors and allow that level an opportunity to resolve the issue. The site engineer for the contractor should meet with the construction management team's resident engineer and discuss the issue using the contract document as a foundation of the discussion.

If the field cannot resolve the issue, it may become a claim, and the contractor must provide written justification from the specifications supporting their point of view regarding the claim to the construction manager. The contractor must also provide a detailed breakdown of all cost- and time-related issues of his claim. The project-level owner's representative, usually the construction manager, should review the contractor's claim presentation documents and request what information is necessary to evaluate the claim properly. Meetings may be held to allow the contractor to present his justification, cost, and schedule information. The construction manager should ask questions and be certain that he understands the claim and all of the information presented. Another meeting may be held to discuss possible resolutions, including acceptance, possible compromise, and/or rejection. Several meetings may be held to discuss compromises. If none can be reached, then the claim is considered irresolvable at the project level and is handed to the next level up the dispute resolution process. Detail should be sufficient at this time to allow the next level in the process to understand the case and meet to resolve the issues.

Meetings at management levels may focus on many issues but should generally remain true to the principles of the contract document. If the top levels of management cannot reach a resolution, then the dispute will be elevated to a dispute-resolution party. This may be a disputes resolution board (DRB), arbitration, mediation, or court as the project documents require.

The purpose of the various levels of the dispute resolution process is to prevent issues from elevating beyond the levels of the parties involved. Disputes that go into the court system are wasteful of time and money for all parties involved.

Dispute resolution meetings are

- held to discuss issues that have caused a dispute between parties to the contract,

- governed by an agenda,

- non-confrontational, and

- professional in tenor and bearing.

Issues not resolvable at one level are passed up to the next level of the dispute resolution process.

(See "Dispute Resolution" section.)

Managing Resources

In one regard, the contractor is like any other business. His business is managing resources. The contractor's resources are manpower, equipment, support tools and supplies, subcontractors, and materials.

The contractor will have bid the project based on estimated production rates of the work crews and the capability and availability of the equipment. Subcontractors and materials are usually lump sum prices that he only manages to maintain schedule and keep changes from happening. If contractor is able to make the estimated production rates with his crews using the tool systems in his hands and the equipment is adequate and stays healthy (available) to do the work, then the contractor will make his estimated profit and be a "jolly fellow."

If, however, for some reason he cannot make the estimates, the attitude on the project site may change. The construction manager needs to have a feel of the pulse of the project to allow him to be aware when these attitude changes are in the air. There are several ways to know when things are becoming problematic for the contractor. We have discussed schedule monitoring, and monitoring of labor burn rate, monitoring of equipment usage. These are key indicators of problems but another and very important method is "management by walking around." The only way to know what is happening for sure is to get out there in the field with the construction crews and the superintendent regularly. With some experience, you will soon realize that the change of tension in the workforce is a palpable feeling.

These tensions are caused by a tension in the leadership. If leadership has problems, very often that tension will find its way to the construction manager as the contractor tries to make back ground that he has lost. Claims, requests for clarification (RFC), and requests for information (RFI) that lead to changes some or all of these begin to show up in the construction manager's in-basket with greater frequency. A weekly trip around the site with the superintendent will do much to defuse these issues and allow the construction manger to help head them off. Adding concepts for success and approaches to the work for consideration—not direction, just ideas—keeps the camaraderie in place as the contractor works through his issues. It goes a long way to keeping the construction manager's in-basket filled with only the normal flow of paper.

The construction manager has a responsibility for the project's outcome, so he has a vested interest in the contractor's success. The contractor is part of a team, and when he has difficulties, the whole team has difficulties. While each member of the team has certain responsibilities, there is no reason that when one member is ailing the others can't pull for him to get well.

The concept of managing resources

- is a feet-on-the-ground concept,

- is preventive "medicine,"

- helps the project as a whole, and

- is a vested interest of the construction manager.

Documentation

Daily Monitoring, and Cooperating with the Contractor

Purpose: To help the construction manager understand that the project in the construction phase belongs to the contractor. The construction manager and his staff are there to see that the project documents are followed during the construction of the project. Moreover, the contractor is not the enemy.

The contractor is the central figure of the project during the construction phase. The result of the contractor's effort is what the whole process is about. The contractor is part of a three-way contract. The contractor signs a contract with the owner, but implicit in the contract is that he must follow the directions of the designer. The directions are, of course, the project documents, including the general and supplemental conditions portion of the specifications, also known as the terms and conditions, the technical specifications, and the drawing set. The responsibility of the construction manager is to see that the contractor completes the project on the timetable set forth in the document (plus or minus time allocated in change orders), completes the project for the dollar amount of his bid (plus or minus money allocated in change orders), and builds what is specified and drawn (including any changes directed by change orders.)

There will be instances when the owner or designer will want the contractor to do something different than directed by the documents. There will be instances when an inspector will find fault with something that is actually acceptable to the project documents. At these times, it is the responsibility of the construction manager to defend the actions and efforts of the contractor.

There will be instances when the contractor doesn't know how to proceed with some protocol, such as how much documentation is required to back up a time and material work invoice. The construction manager and his staff may need to spend some time training the contractor.

The key to working successfully with contractors is to be firm but fair. Give the contractor his due by contract and allow him to control the building of the project. Tools should have been set in place for the construction manager to take control of the project, but only if the contractor abdicates that role by breaching the contract responsibilities as spelled out in several places in the general and supplemental condition sections of the specifications.

For the very reasons stated above, the construction manager and his staff must know the contract document better than anyone associated with the project.

The process of daily monitoring and cooperating with the contractor

- helps the work flow,

- keeps the construction manager in touch with the project, and

- presents a teaming atmosphere on the project work site.

Communications

Purpose: Keep communication lines open.

Communication between the various stakeholders of a construction project is as key as it is among the troops on a battlefield. When the commander loses communication on the battlefield, he can no longer expect to have the initiative, and the advance of his troops stops.

On a construction project, if the stakeholders stop communicating, the bickering begins and the project slows down due to lack of command. Breakdowns in communication usually happen because one party or the other feels that the contract is being misused in some way. A firm interpretation of the contract is important as long as the term fair is used in the same sentence. At the outset of the project, the construction manager needs to let all of the parties know that the contract will be interpreted in favor of the contract, not any of the parties.

Never close the lines of communication. No matter how surly other parties become, maintain the professional decorum of a manager. When writing tough letters, do not allow personal feelings to corrupt your professional ethic. Let the contract dictate the language. Do not attack individuals, attack issues. Even if the construction manager reaches an impasse and cannot compromise to an agreement on pricing of a request for proposal (RFP) he must stand firm, agree to disagree and then move on rather than lose his professionalism. Very often just maintaining your decorum will help keep the disputing parties at the table closing in on a resolution.

The concept of partnering was developed to establish relationships between the stakeholders of a project so they would feel comfortable communicating with each other in an amiable fashion and to develop a protocol for resolution should an issue reach an impasse at any level. (See "Dispute Resolution" section.)

Very often if the construction manager is concerned about an issue and he doesn't know how to approach it, he can go out and exercise a little verbal initiative with the contractor's designee opposite him on a friendly, nonthreatening, off-the-record basis. He should get some feedback about his concern. The issue may not be as big a deal as he first thought, or it may be a sore that needs a little more than the first aid that he thought at first.

If there is an issue that needs to be brought to the table, resolve it while it is fresh. Never let sores fester or "a sleeping dog lie." Festering sores have a bad way of getting infected and going from sore to cancerous in a hurry, and sleeping dogs eventually wake up. Usually what went to sleep as a churlish Chihuahua wakes up a ripping Rottweiler.

Open communications

- prevent issues from getting out of hand,

- keep the parties on an amicable basis, and

- keep issues from festering.

Monitoring of Local Landowner Agreements

Purpose: To give direction to the construction management staff regarding their responsibility to oversee the proper treatment of the property on which projects are constructed.

All of the appropriate land ownership, easement, and right-of way documents that define the construction activities by the contractor while he is working on the land owner's property should be included in the construction bid documents as requirements for conducting their construction activities while working on a given property. Each party must be careful to follow the rules laid out in the contracts.

Any violation of these contracts may be considered breech of the contract with the owner and will usually require immediate remediation. Immediate action to rectify misuse of property by the contractor will maintain a high level of good relationship between the land owners and the project owner. It is important to maintain that relationship in good order. The construction manager might invite stakeholder agencies or land agents to progress or special meetings to voice concerns or potential pitfalls to the construction team.

As construction progresses, the contractor should be required to provide a three-week look-ahead schedule that defines the activities the contractor is planning to conduct over the upcoming three weeks. If the contractor is constructing pipeline within a utility corridor, the contractor should be required to define those activities by pipeline station so that the construction management staff can keep abreast of progress in landowner-affected areas.

Monitoring of local landowner agreements, rules, and regulations:

- Include all special activities, rules, and requirements for local landowners in the construction documents.

- The three-week look-ahead schedule should include movements from one landowner to another so the construction manager can monitor requirements.

Monitoring Project-Acquired Permits

The contractor may be responsible for obtaining most permits for construction of the project. Those will include any water discharge permits for dewatering operations, general use permits and stripping permits, clean air permits, special use permits, and building permits. It is the responsibility of the construction manager to monitor the acquisition of these permits to make sure the contractor does indeed obtain all permits that he is required to obtain and that they are obtained in a timely manner before the work for which they are needed is started. If the contractor is delinquent, the authorizing government agency may have issues that create all kinds of headaches that could cause slowdowns or even stoppages of the work progress, locally or across an entire program.

The same considerations apply to all environmental regulations and landowner rules and stipulations. The contractor needs to know what he is facing in both of these areas to ensure that obedience to both is complete. Often environmental permits are based on compliance. Noncompliance may bring stiff penalties or work stoppages until all issues are rectified or mitigated to the satisfaction of the issuing agency or authority. In green areas, scheduling of portions of the work may be driven by environmental issues. The construction manager must be fully aware of environmental drivers and work with the contractor and environmental agencies to stay within the guidelines set in the permits.

Monitoring project-acquired permits and land:

- Include all special activities, rules, and requirements for permits in the construction documents.

- The three-week look-ahead schedule should include movements from one permitted area to another so the construction manager can monitor requirements.

Reports

Inspection Reports

Purpose: To introduce the concept of keeping an accurate daily journal or inspection report.

The daily inspection report is a record of daily activities

- performed by the contractor.

- that the inspector accomplished in observation of work performed by the contractor.

- performed by the inspector as he or she interfaced with the contractor, designer, owner employees, and/or other individuals that come to the project site.

An accurate record of the date, day, general weather, temperatures, important people on the site, including him/herself, visitors, contractor's management, owner's personnel, designer personnel. Contractor's scheduled activities (as scheduled on the project construction schedule), observations, verbal or written correspondence, such as direction to correct a defect and result, suggestions, and/or assistance given to the contractor should all be included on the daily report.

Inspection is done not to find problems and whip the contractor with them; it is to prevent problems from being incorporated into the work and ultimately the finished product. If a problem is found, notify the contractor's foreman and give the contractor a chance to correct the problem. If the problem is corrected, note it in the daily report. If not, work with the contractor to get the problem fixed before it becomes a permanent part of the project. If the contractor refuses to address the problem, note it, take record photographs, and notify a superior. If the foreman refuses to act, go to the superintendent. If the superintendent refuses to act, go to the construction manager with the problem before the problem is incorporated into the finished product. If after all you can do the problem becomes permanent, note the problem and fill out a nonconformance work report, attach photographs, log the report, and give the report to the contractor.

Be sure to sign and date every sheet on which you have written any part of your report.

Each inspector will fill out a daily inspection report. All reports are discoverable for court actions as evidence, so keep them clean, with no erasures, no foul language, and no innuendo, only facts based on observations. All activities in progress, including start date and/

or completion date, will be tracked on the project construction schedule.

Inspections

- are not punitive, vindictive, or self-protective.

- are to discover items improperly prepared for construction and get them corrected before they are constructed.

- are to give the owner the best constructed project possible.

Construction Manager's Report

Purpose: To explain the use and purpose of the construction manager's report

The construction manager's report is an audit report based on the reports of the inspectors under his direct supervision and a report of his interface with the contractor's personnel, and others with whom he may have had a pertinent conversation about the project.

As a course of business, the construction manager should periodically walk the site looking at various items to see what is happening, observe means and methods, count craft, and generally observe the practices of all personnel on the site. The construction manager periodically (at least weekly) should make this walk through with the contractor's superintendent or project manager and discuss what is observed. Shared activities build trust and confidence that the construction management staff care about the contractor's work and enhance communication between the construction manager and the contractor. Point out successful activities with pride just as you may rebuke the contractor for problems. These project walks are a good time to discuss off the record any issues that may be of concern to you or the contractor.

In the monthly report, note highlights of meetings that are of particular concern and refer to the minutes of the meetings for detail. Note issues that need follow up and then note on the future report what action was taken.

All reports are discoverable for court action as evidence, so keep them clean, with no erasures, no foul language, and no innuendo—only facts or opinions based on observations.

The construction manager's report

- documents contacts.

- issues general progress on the construction project.

- should be made following a periodic walk of the project.

Construction Manager's Monthly Report

Purpose: To define the requirements of the monthly progress report to the owner. The form of this report should be coordinated with the owner and project manager.

The monthly report is a project status report. Each month the construction manager will prepare a report that is submitted to the owner's representative. The report will contain all pertinent information that represents the current status and progress of the project. It also includes the original value of the project, the current value of the project, the original completion date and duration of the project, and the current completion date and current duration. It includes total change orders to date and the change orders that have been submitted during the last month. The current change orders will be spelled out individually by definition, value, time, and cause. A list of known issues pending and possible ramifications to the project, if any, are stated. A text describing the physical progress during the past month will be written, and an estimate of the state of the project in physical completion, schedule status, and cash flow terms will be given. The construction manager will also project the next month's activities.

This report may be combined with the reports from other projects. The owner's representative may summarize the construction manager's reports from all projects into a total construction phase report. This total report or just your construction manager's report may be reported at higher levels within the owner's organization.

The construction manager's monthly report

- is a report of the construction progress accomplished on his project over the past month.

- should include dollars spent during the past month, along with dollars spent to date.

- may be combined with other project reports for the past month.

Weather Reporting

Purpose: To explain the importance of daily weather reporting.

Most contracts have a weather time and delay clause. This clause allows the contractor to claim weather delay if there are more than an allotted number of days that he cannot work because of weather conditions. This clause may also reference and define weather norms that allow contractor to claim weather delay when abnormal weather impacts his work. The daily weather section on the daily inspection report is the measure for claiming those delays. Accurately reporting the daily high and low temperatures, the general weather condition, and the amount of moisture that fell, are all important to protect the owner from weather delay claims and the contractor so he may safely receive credit for all appropriate weather delays due him under the contract. Weather delays may only be claimed when critical path activities are impacted.

Daily weather report

- is the basis for awarding weather-related delays.

- should report on all weather activities that might be construed as delaying the contractor's ability to perform work on the *critical path activities* of the construction schedule.

Monitoring of Construction Documents

Construction Schedule

Purpose: The project construction schedule is the prime tool of the construction manager for ensuring that the contractor is in control of the project. Because it is such an important tool, schedule maintenance is tied to breach of contract. The project construction schedule is the contractor's plan for building the project. Many contractors don't realize that, but it is a fact. If the contractor gives the schedule proper consideration during the estimating process and then uses the schedule to guide the project, he will generally lead the project to a successful completion.

The project construction schedule needs to be reviewed to see that it includes all required milestones and that it has been developed to include all important activities and tasks, such as submittal review time, manufacturing time, and delivery of key equipment/ materials to a sufficient detail so it demonstrates the contractor's understanding of the project and can be used to track progress or lack thereof as the project goes along. The logic of the sequence and duration of activities must also be scrutinized in some detail.

The construction manager or a master scheduler should review the initial construction schedule to make sure that all of the major activities, critical milestones and other critical dates are properly handled and accounted for. He should also check logic and logic ties for accuracy and efficacy.

The contractor should be required to provide an up-dated project construction schedule with his periodic (monthly) payment application. The construction manager should review the monthly up-dates to make sure that contractor is reporting his progress accurately and that all change orders affecting contract time are incorporated . The construction manager must also be sure that any slippages, accelerations, and change orders are reported and properly reflected in the update. Changes in the critical path are extremely important. If slippages occur, a corrective or "recovery" schedule may be required. See below in schedule remediation section.

The contractor is also required to provide a three-week look-ahead schedule at least twenty-four hours prior to the weekly progress meeting. This small window or "snapshot" schedule should be compared to the project schedule to determine immediate scheduling issues (see "Contractor's Short-Term Look-Ahead Schedule"). The construction manager should compare that weekly snapshot with the construction schedule to see if there are indications of slippage.

The construction manager should review the construction schedule to ensure that it

- includes all critical milestones, dates and activities.

- is properly updated including activity slippages or accelerations.

- Is updated monthly or at every pay request and at every change order.

Submittals

Purpose: To develop an understanding of the specifications requirement for checking on the contractor's choice of materials and permanent equipment to be used on the project.

The construction contract has requirements that generally cover the procedures for submitting general information and data regarding critical materials and equipment to be installed on the project. The contractor should read the submittal requirements and each section of the technical specification section of the project manual to

1. understand how to review a vendor submittal,
2. understand how to submit on material and/or equipment, and
3. learn which material and equipment require submittals.

After he has learned which materials require submittals, he needs to review the project schedule and compare the list of items requiring submittals to the project schedule so he knows the earliest date that each item will be needed.

The contractor should make a submittal schedule listing each item requiring a submittal followed by a date that the item will be submitted, allowing ample time for processing through the system and full contract time for at least one review before the "early needed" date. This submittal schedule is then submitted to the construction manager for review and upon acceptance is used by the contractor, construction manager, and designer to plan the submittal review process.

The construction manager's staff should also carefully read the specifications and determine what submittals are required and generate a submittal log as a check on the contractor's submittal log. The contractor is required to submit a submittal schedule so the construction manager knows that contractor is aware of all of the items he is required to submit on. Any discrepancy between the contractor's submittal schedule and the construction manager's submittal log should be explored and remedied.

The contractor is responsible to review his submittals prior to submitting them to the construction manager for designer review. The lack of review by the contractor causes more rejections than any other single reason. Multiple reviews are inevitable but should be rare and generally due to misunderstanding of the specification on the part of the contractor or near equal qualifications for equipment/devices or material. Frequent rejections and/or amend and resubmit comments are a red flag that contractor is not following the specifications by reviewing submittals prior to submitting them. Some contracts stipulate only a specified number of reviews of the same submittal are allowed. After that number, the contractor will be charged a set dollar amount for each re-review. These amounts are normally deducted from the contractor's next payment application

For complex items or items that may not be as clearly defined as they could be, submitting early enough to allow for a second review may be important. The project document may not allow for more than two submittal reviews. The contractor may be charged for each additional review beyond two reviews allowed. The contractor may also find that multiple reviews may adversely affect his construction schedule.

The construction manager should monitor the submittal schedule carefully and look very closely if submittals begin to slip or require multiple reviews. The submittal log will be copied to the contractor and the designer at each weekly progress meeting for them to compare with their own log. Discrepancies will be brought to the attention of the construction manager immediately so that all parties are coordinated.

All of the following should apply to project submittals:

- The construction manager should know what submittals are required.

- The construction manager should require the contractor to submit a list of submittals and a schedule demonstrating when he will submit each submittal.

- The construction manager should review each submittal before he sends it on to the designer.

- The construction manager will maintain a submittal log.

Requests for Information

Purpose: To explain the question asking and answering process.

The contractor will no doubt have questions during the course of each project. Some things that seem perfectly clear during the design phase are not clear at all when viewed later in the context of the entire design. An avenue is available for asking questions about the design and/or written specifications, their meaning, and/or their intent. That avenue is the request for information (RFI).

RFIs are usually initiated by the contractor on a provided form. However, the construction manager or owner can submit an RFI to the designer using the same form. The RFIs are sent to the construction manager for review and logging. The construction manager should be prepared to route them to their proper destination, which is most often to the designer. Some will stay in house, and a few may go to the owner. RFIs are usually time sensitive and should spend as little time as possible in the construction manager's hands.

The contractor usually sends an RFI out as soon as he has a question. There should be a minimum of three weeks before an answer is needed. However, that is not always the case. Occasionally the contractor will ask a question as a result of preparing a submittal, the vendor will send an RFI through before he submits on his product, or the field superintendent will need a question answered before an activity can begin. So the submittal, product order, or activity may be delayed until the answer to the RFI is returned. Usually the designer has a certain amount of time defined in the project document to answer an RFI that begins the day the construction manager receives it.

The construction manager will make and maintain a log of submitted RFIs and track them from receipt to return to the contractor. A copy of each RFI will be made when it is submitted and when the answer is returned to the construction manager from the designer. A copy will be in each project's work space so that all inspectors, administrators, and designers know what the answer is. Often times RFIs lead to changes that require a request for pricing (RFP) to be issued to the contractor. That is a third reason for quick turnaround times for RFIs.

If the construction management staff has a good relationship and daily conversations with the contractor, he may ask a question in the field that requires an RFI. However, the construction manager can start the ball rolling on critical issues by issuing a "heads up" memo with the contractor's question to the designer so that he can get a jump on researching an answer. The heads up will only predate the RFI, not replace it, and time starts with the RFI issuance to the construction manager. A heads up is particularly important when it appears that some additional design or design change may be required and/or a schedule impact may be imminent.

The RFI log will be copied to the contractor and the designer at each weekly progress meeting for them to compare with their own log. Discrepancies will be brought to the attention of the construction manager immediately so that all parties are coordinated and questions do not get lost in the shuffle of paper required to keep a project running. The construction manager will watch the time of possession for each RFI and serve as a "tickler" to the designer to remind him when an RFI is nearing its contract time in review. Changes to the design, whether by work change directive (WCD) or field order (FO), no matter how minor, caused by RFIs will be shown on the as-built (red line) drawings.

RFIs

- are the contractor's method of asking questions.

- are available to all parties to ask question.,

- require swift action because they may affect the start of an activity

- require swift action because they may be holding up a submittal.

- are possibly the first step to other actions, such as change orders.

- are reviewed by the construction manager coming from the contractor and returning from the designer.

- are important documents that must be transferred through the system as rapidly as possible.

- are logged and tracked by the construction manager.

Design Clarifications

Purpose: Not all questions are answered with an RFI. Sometimes a clarification of the design or specification is all that is needed.

Design clarifications are used more by architects than engineering disciplines. Requests for clarification (RFC) are asking for a clarification on a drawing because the intent or information given is not clear. An example might be a dimension that does not print well or a note requiring all of the walls in an area to be of one material except one wall undetermined. These types of questions will usually be answered by a clarification. They differ from an RFI answer in that the ramifications to a clarification are very limited, if any. They are usually verification to the contractor of what he thought was the case but had enough doubt about to ask a question. A clarification will very rarely cause a design, cost, or time change to the project.

Clarifications are to be tracked on a design clarification log. The clarification form and log usually is in the same format as the RFI form and log.

This similar referencing is extremely important and points to the need for consistency and continuity in all of the documentation of the project.

Design clarifications

- are questions that clarify the design drawings and

- are tracked by the construction manager on a design clarification log.

Correspondence Log

Purpose: To provide a tracking mechanism for all correspondence from and to the various parties sharing ownership in the construction phase of the project.

The owner, designer, contractor, and construction manager are the primary parties that are involved in the construction phase of the project. However, there are many other parties that will have some part in providing equipment, parts and pieces, and/or effort in the construction phase. The lines of communication are well defined and must be maintained according to contract guidelines. That is not to say that communication will not breach the contract boundaries. Therefore, a set of files defining the various relationships will be set up to keep all correspondence together and intact. A communication log will also be established to track all communications not being tracked on the submittal log, the RFI log, the clarification log, the RFP log, or the change order log. The construction manager can create a log that fits the best format for the project or company policy.

Letters, memos, and emails are of particular importance. Letters will be tracked in a log. They will be logged and tracked by number.

All correspondence

- should be logged and saved and

- should be tracked by the construction manager on a correspondence log.

Applications for Payment
(See Resource Loaded Schedule
and Schedule of Values)

Purpose: Progress payment handling procedure.

The contractor may be required to submit his pay requests on a project-specific form when applying for any payment, be it a regular progress payment, a project completion final payment, or a termination payment. The request for payment format may be one developed by the owner. The most popular is the standard American Institute of Architects (AIA) form for a cover sheet with a spreadsheet backup, which contains the schedule of values. The owner or construction manager may provide this form.

A construction management staff member meets with the contractor's designee to discuss and reach agreement on the periodic (monthly) invoice. The contractor takes the agreed-upon numbers and submits an invoice to the construction manager's staff member. If all is in order, then the construction management staff member recommends the invoice and gives it to the construction manager for review and comment. If there is an issue, the construction manager will deal with it. If not, the project administrator uses the billing as the basis of the monthly project construction report, copies it for the file, makes a cover letter for the construction manager's signature, and sends it to the project controls office for payment.

Final project payment requests must be accompanied by a checklist (punch list) of all of the items required before final payment can be made. The checklist must be filled out by the contractor and initialed by each item by the resident engineer before final payment can be recommended for payment. Request for retention must follow the required procedures and can only be requested as the project document allows and/or after final payment is made. The checklist is project-, owner-, and regulation-related and must be generated specifically for the project.

Payment for termination may only be made after the resident engineer has reviewed the terms of the termination and determined that they are met and the project counsel has also reviewed the termination agreement and payment for accurate reconciliation. Then the payment will be processed as a final payment and a report regarding the termination including justification, agreement, and recommendation for contractor disposition will be attached to the recommendation for payment.

The contractor is due a swift turnaround on all payment issues. The contractor's results are the entire reason for the project in the first place. Treat the contractor fairly. The contractor has no responsibility to fund the owner's project. There may be state or local statutes that regulate payment and/or retainage. Be sure to research and abide by these statutes.

All of the following apply to payment applications:

- Requests for payment should be reviewed by the construction manager and recommended to the owner for payment.

- All closeout items must be checked before final payment can be recommended by the construction manager.

- Payment applications should be processed to the owner for payment as quickly as possible.

- A thorough review of the project is done and a list of work to be completed is kept.

- Create a punch list and attach it to the certificate of substantial completion.

Training of Owner's Staff

Receipt and Distribution of Operation and Maintenance Manuals for Comment

Purpose: To provide an understanding regarding Operation and Maintenance manuals (O&M manuals) and their importance.

The contract document requires operation and maintenance manuals (O and M manuals) be provided for every piece of significant equipment and some systems. These manuals are the directions for the plant/system operations and maintenance staff. Therefore, it is important that contractor be required to provide enough manuals at submittal to allow both the appropriate design staff and the operations and maintenance staff to review and comment on the submittal. In fact, it is recommended that a full-day workshop is planned to coordinate the review and comments.

The workshop should be held a week or more after the O and M manual submittals have been distributed to allow all of the participants to have an opportunity for review. The O and M manuals must be in the contractor's hands before equipment assembly and/ or setting begins. In fact, the contractor should have the O and M manual for each piece of equipment before the anchor bolts are set. Take advantage of these early copies to get the designer and owner's staff up to speed on what is coming their way.

Operation and maintenance manuals

- should be developed by the engineers who designed the operations process,

- should contain operation and maintenance information for every piece of equipment in the process and support processes, and

- should be reviewed by the construction manager and the owner's operations staff.

Review Comments

Purpose: To provide an understanding regarding submittals and their importance.

Prior to returning any submittal to the contractor for modification and resubmittal, the construction manager should pull one marked-up set for his review file. The construction manager must review the comments for comparison with the resubmittal to make sure that all comments are addressed. Resubmittals must also be distributed to the designer and the owner's staff for review and comment. When both reviewing parties are satisfied that all pertinent information is in the manual, then the designer can stamp the submittal as approved. The construction manager must retain a copy of each submittal and resubmittal, both plain and marked up. These retained copies are the record of submittal quantities, a record of the thoroughness of the contractor, and a history of the thought process used to, in many cases, to evaluate the vendor's efforts toward the project and his equipment.

Training by Manufacturer's Representative

Purpose: To set out guidelines for evaluating vendor training.

In the specifications, various equipment vendors will be required to train the operations staff regarding the operation and maintenance of their particular equipment. Read the specification carefully. The requirements may vary.

Some may be required to bring a working model, others a video and critical parts or pieces to use for demonstration purposes, and still others an O and M manual and verbal presentation. Whatever is required, **be sure that the manufacturer's trainer is *not a salesman***. The trainer must be from the design, maintenance, or development side of the manufacturer's operation. The trainer must be well versed in the operation, troubleshooting, and maintenance of the equipment on which he is training. Time must be provided for hands-on demonstration and a lengthy question-and-answer session.

Chapter 5

Contract Changes

Purpose: To establish a process for obtaining cost and schedule impact for a change to the contract.

Contract changes can come at the request of the owner, designer, construction manager, or contractor. Generally, the order of a change to the contract comes as follows:

1. Most often they come as the result of a question asked by the contractor in a request for information (RFI) document in which the contractor asks for information that he cannot find in the project documents or a discrepancy between pieces of information in the documents. These sometimes result in a design modification in answer the contractor's question. The contractor may request a change to the contract or disagree that an activity has no cost or time impact. In that case, the contractor may submit a request for change, (RFC) or in some documents, this action initiates a claim.

2. In response to the RFI or RFC, the designer may institute a change in the design for any number of reasons, or the owner may request the contractor to perform some activity, either of which are routed to the construction manager.

3. If the design change or work to be performed will change the cost and/or time of the contract, the construction manager will request that the contractor evaluate the change and provide a cost/time estimate. That request is called a request for pricing or proposal (RFP).

4. The contractor will estimate the change and respond, providing a detailed estimate with backup.

5. After review, possible negotiation, and agreement on the price and/or time of the RFP, the construction manager will write a work change directive (WCD) (see work change directive section) if there is an impact to either time or cost or both.

6. If a change is initiated for which there is no perceived cost or time impact, a field order (FO) is issued.

7. Change orders are the only instrument by which the contract may be modified. WCDs are included in change orders to define the cost and/or time modifications. FOs are included to define no-cost items and/or time modifications. Claims initiated by the contractor and approved by the owner are included as time and/or cost modifications that did not follow the above six-step process. They are defined below.

See section requests for information (RFIs). Requests for Change and Claims will be discussed in detail later.

Request for Pricing (RFP)

If the contractor deems some direction or request as being a change in the scope of work, he can send a request for change order (RFC) (see request for change section) to the construction manager. The construction manager will review the RFC, and if there is justification for a contract change, he will respond with an RFP. Once the RFP is returned, in all cases it is handled as follows:

- The contractor has a period of time—usually ten days—to respond with an estimated price and schedule impact. The response must be in detail, with pricing broken down for labor, equipment, and material, with the contract-allowed markups applied as a separate line item. The applicable taxes, bond and insurance cost is then added to the subtotal of costs to reach a price of the RFP. The RFP is reviewed for merit and entitlement if originating from contractor demand, accuracy, and completeness. If the RFP is returned with insufficient detail, the construction manager will send it back with a request for more detail of whichever area is lacking. Minimum detail as follows:

- unit pricing for each item with total units, unit cost, and extended cost (These items are all entered at bare cost. Markups are to be applied in a subsequent step.)

- subcontractor allowed markup

- general contractor allowed markup

- subtotal of costs

- applicable tax on material or equipment. (Note that tax may not be applicable to some clients if purchased freight on board (FOB) jobsite. This treatment usually applies to government entities, such as cities, or quasi-government entities, such as school districts or special districts.)

- bond and insurance adders on base price

(NOTE: Add these costs in this order so that mark-up or margin is not applied to taxes or bond and insurance costs.)

When sufficient information is received by the construction manager, it will be evaluated against a standard, such as known costs from in-house sources, RS Means estimating manuals, Richardson Cost Manuals, or other accepted standards that identify cost per unit of work performed. Charge rates for construction equipment are compared with a standard rate as defined in the project documents, such as Blue Book or State DOT rental rates, and material prices are compared against known pricing of the same or similar items. Consideration may be given for local market conditions, labor, material, or equipment availability and cost to ship or mobilize needed goods, support equipment, or labor.

If the pricing is in order and if more or less time is needed to accomplish the change in work because the change affects the critical path of the project construction schedule is acceptable, then the RFP is accepted and a WCD is issued in the amount of the agreed-upon price and the time is added or subtracted to reflect the calendar days added or subtracted. If the construction manager does not agree with the pricing and/or time allocation, he should meet with the contractor to go over the detail of the RFP to see if both he and the contractor understand everything in the proposed change. If they agree on the scope, then the construction manager should review the pricing and explain the differences to the contractor and ask the contractor to review his pricing and/or schedule impact. The construction manager may have to do some review as well or may have misunderstood something and have to review his information again. This process may take several iterations. Once the RFP is accepted, the construction manager issues acceptance of the RFP to the contractor.

Usually the designer will tell the construction manager when an RFP is warranted, but not always. It may fall to the construction manager to determine if the change will have a cost and/or time impact. If the design change results in a cost or time impact, the construction manager issues an RFP to the contractor. For example:

Simply moving a wall two-inches from a given location may have no time or cost impact if it is yet to be constructed. If this is the case, a field Order (FO) (see field order section) would suffice. However, adding a doorway and a door to that wall would constitute a cost impact and possibly a time impact whether the wall is in place or not and would require an RFP, resulting in a WCD. A constructed wall that must be moved because of redesign would definitely have a cost impact and possibly a time impact. Time impact allowances are only given to activities on the critical path that affect the overall schedule. Some items with time impact may be side activities until they are modified and are time impacted to a point that they are pushed onto the critical path.

If the construction manager reviews, checks, and accepts the price and required time adjustment as equitable and fair, he will recommend it for acceptance to owner's representative depending on the project's organization structure. If the price seems exorbitant and/or the time requested is not fair to the owner, based on the construction manager's staff-prepared estimate of cost and time, the construction manager will open negotiations with the contractor, and they will negotiate in good faith to an equitable agreement. If no equitable agreement can be reached on the issue of cost and/or time, the construction manager will recommend that a change not be given to the contractor and will recommend that other options be pursued such as bringing in others to accomplish the work in question. This course must be weighed carefully because bringing in others, whether they be the owner's people or another contractor, to accomplish a piece of the work on the site could cause larger impacts to the overall project, such as impacting the contractor and causing a delay to his work.

If agreement is reached, the construction manager will issue a WCD recommending it for owner's approval. If the owner gives approval at all required levels, the WCD becomes part of a change order that is signed and issued. When the change order is issued, the contractor must add the changes to his project construction schedule, the resource loaded schedule, and red-line drawings.

Summary of requests for pricing:

- An RFP is the device through which the construction manager obtains a proposed cost for proposed change to the contract.

- RFPs should have a parallel check estimate developed by the construction manager.

- RFPs are the bases for negotiations for proposed changes to the contract.

- If /when agreed pricing is reached, the construction manager can recommend a work change directive.

Field Changes

Documented in Requests for Information

Purpose: To lend understanding to the field change process and the responsibility of the various parties involved in the project.

Field changes are not made at the discretion of the field staff. The designer of record has full responsibility and liability for the design. Added features, elements, or scope changes may be made in the field at the direction of the owner through the designer to the construction manager and on to the contractor.

All field directives must have documents to justify the directive. RFIs, as we have said before, are the most common origin of field changes. However, not all RFI answers have a cost or time impact. Some make a minor change or even a major change that was able to be included in the work replacing or moving some design element at no cost. These are called field orders (FO). The only documentation directing these sorts of field changes is the RFI and the resulting FO.

The FOs are listed as no-cost items in the various change orders. It is required that the changes made through the FOs will be noted and included in the "red-line" as-built drawings by the contractor. Once in a while, a single one of these is of large enough scope or important enough that it will generate a no-cost change order. RFIs that have a cost or schedule impact are documented to the contractor as first RFPs, and then if accepted, as WCDs. If the work is to begin immediately and there is no time or not sufficient time for the contractor to prepare an RFP or if the RFP is not accepted in time for work to begin, the construction manager issues a WCD to the contractor. He will also issue an RFP or a directive to begin tracking all cost-related tasks and materials that relate to the WCD. Under a change order, the construction manager will require that the work be tracked on a time and material basis until such time as an agreement is reached regarding the cost and time impact.

The owner may request changes by adding elements or features that add scope to the project. These issues will all be routed through the designer for design implementation and then sent to the construction manager for field implementation.

As with all field changes, these shall be incorporated in the red-line as-built drawings as required by the contract document for the contractor, which should be reviewed periodically and submitted at the end of the project.

Field changes are

- documented in RFIs, RFPs, WCDs, FOs, and ultimately change orders.

Work Change Directives (WCDs)

Purpose: To allow a contractor to begin work on an approved change activity that has not been compiled into a change order

A work change directive (WCD) is an intermediate document that directs the contractor to go to work on an activity that is a change to the document before a change order is issued, knowing that activity has been approved and the owner is obligated to pay for the performance of the activity. Most project documents require that a contractor not work on a change until an approved change order is in place. However, some changes require immediate action or will take a while to go through the process of formal acceptance and must be completed before the process can run its course. If every change was accompanied by a change order, there would be many small change orders on a project. Most projects set a threshold for change orders of some amount. Legitimate changes smaller than that cannot be ignored, so we go through the process of getting them approved for payment by issuing a WCD directing the contractor to do the work at an approved price on a time and material basis and/or with an approved schedule change. Then the WCD may be held until there are several that add up to the threshold amount. When the threshold amount is reached, a change order is written including all of the WCDs and/or field order items accepted and outstanding.

If an RFP has been reviewed and accepted at the field level and verbal acceptance can be obtained at higher levels but formal approval may take some time and would delay the work, a WCD can be issued that obligates the owner to pay so that the work can continue on schedule. Another appropriate use of WCDs is to move forward on change activities requiring immediate attention, such as emergencies, or activities that, if delayed, would cause schedule slippage. This is done by authorizing a WCD with an accompanying RFP on a time and material basis where all labor, equipment usage, and materials are tracked in detail and approved on a daily basis. At the end of the activity, cost and time are verified by the approved daily tickets.

For Work the Contractor Has Constructed

If the parties at the project level cannot agree and the proposal of the contractor is rejected, the contractor may accept the construction manager's evaluation or the parties may resort to a WCD on a time and material (T and M) cost basis. The T and M process requires contractor to produce a summary of work and costs by labor and equipment type for signature of the observing construction management staff person on a daily basis. If the work has been completed, with no resolution having been reached and no record of actual time and costs having been recorded, the contractor has done so at his own risk and may file a claim, which then goes into the dispute resolution process. It is the construction manager's job to keep that from happening.

For Work Not Yet Constructed

The construction manager may pursue alternate means for construction other than the contractor if he deems that the proposed price and/or time requested are not fairly offered. But the construction manager must evaluate impacts to the contractor's ongoing work efforts on the balance of the project.

When submitting a work change directive, the construction manager must provide sufficient information to verify that indeed each work change directive is

- justified,

- properly priced (except FOs),

- properly negotiated as necessary, and

- tallied against the standard of care agreement if one has been negotiated on the project.

A chain of paper, including the RFI or RFC, RFP, WCD, and backup are included.

Work change directives

- are intermediate documents between requests for change and formal change orders, and

- are supported by initiating documentation, such as an RFI, RFP, negotiated pricing, or time and material directive.

Field Orders (FO)

Purpose: to identify items of scope, workaround issues, or contract language that changes the contract but has zero cost and/or schedule impact to the contract.

Field orders are a change to the work or contract language that has zero cost impact and no time impact to the project. An example would be modifying a wall with a door. Adding a door would require some cost and time, but moving the door already designed in the wall may require neither cost nor time change to the contract. These types of changes happen all of the time from the first day to the last day of a contract. Field orders are markers that document these minor changes and provide some history of changes made that have very little paper to document them. Field orders are included in change orders as zero cost and no time change, but once again, the documentation is there that something happened and someone in authority approved it.

- Field Orders are changes to the construction contract document that have no cost or time impact.

Change Orders

Purpose: To provide a controlled means whereby the contract can be altered, adding or deleting scope, time, and/or dollars.

Contract change orders are the only document that can modify the contract documents. Modifications to scope of work, dollar amount, and/or the time allocation as described in the construction document are made by change orders. That protocol makes the change order the second most-important document, second only to the contract document itself. When incorporated, the change order becomes part of the contract. The change order consists of all outstanding WCDs and FOs that have been approved. The construction manager must keep a running log of all change orders. This log will contain all WCDs, FOs, and RFPs (RFPs are pending WCDs) and their dollar amounts and time allocations. WCDs and FOs are totaled and summarized in the change order document.

Change orders are constructed on a contract change order form. The change order document will require a space for "change order description." In this space, the included WCDs and FOs are listed by number and a brief description of what the cost and time of each represents. The value of each is listed on the right-hand side in columnar format. The total of the column must match the total value of the change order.

The total value of the change order is added to the most recent contract value. That is to say the value of change order number one will be added to the original contract amount. The value of change order number two will be added to the new total of the original contract amount plus change order number one and so forth until at the end of the project, the adjusted contract amount equals the total of the original contract value plus the collective values of all of the change orders.

Each change order will consist of the following:

1. The cover sheet and the completed change order form.
2. A copy of each WCD and FO is included in the change order and all of the backup for each WCD and FO. Each WCD and FO is its own packet.
3. Finalized claims will be included along with all backup for costs and time allocation.

There are places for three signatures on the bottom of the cover sheet. They are for the following:

1. The contractor
2. The construction manager
3. The owner

In order for a change order to be executed and in effect, all three signatures must be in place.

Copies of each change order are made for original signatures according to the requirements of the project. At a minimum, these are distributed to the following:

1. The owner's representative
2. The owner's controller, chief financial officer, and/or head accountant.
3. The contractor
4. The construction manager
5. The designer of record

Time added or subtracted is noted in the change order, and a new total contract time is established. The contractor must, from that point forward, account for time according to the new total allowed contract cost and time.

The construction manager may be asked to track cause of change orders, where change orders are a composite of many WCDs and claims each part will be assigned a cause. Typical causes for changes to the contract include claims against the designer that may be caused by

- owner-directed changes,

- differing site conditions,

- designer error,

- designer omission,

- back charges to the contractor

A change order also triggers a requirement in the contract for a new construction schedule to be generated. The new schedule will be adjusted to reflect the added or subtracted time allocation along the critical path of the schedule. Unaffected construction activities should not be changed.

- WCDs are issues proposed as changes to the contract that have cost and/or time implications to the contract. They can be based on hard costs derived from a contractor-priced RFP or in the case of emergencies, used to direct execution of time and material work that is closely tracked or if the timing of the work is critical and a price cannot be agreed upon or is in negotiation can be used to direct work forward with cost and time being tracked as T and M until an agreeable price is reached. In the last case, if none is reached, the T and M price and time may be used to develop a change order.

- A field order (FO) is a proposed change to the contract that has zero cost and/or time implication to the construction document

- *A change order is the only method by which the contract can be altered; therefore, a change order might contain one or more WCDs and/or FOs.*

Request for Change Order (RFC)

Purpose: To define a process for the contractor to address situations wherein he has been requested to do work or has completed work for which payment has not been offered.

The contractor needs a method to request a time and/or monetary change to the contract that he feels he has earned. Often the issue is that the contractor performed an activity that he did not know was a requirement of the document or he missed something in his bid. Neither of these are legitimate reasons for a change order. However, there are issues that arise wherein the contractor has a legitimate claim for extra time and/or monies to cover an activity or materials requested by the designer and/or owner. These usually come to light initially as an RFI. The designer responds to the RFI directing something more than the document requires or a change that requires more cost than the contractor could have reasonably expected when he bid the project. If the designer and/or construction manager do not recognize the added cost or time requirements, the contractor may issue an RFC for the construction manager and designer to consider. The construction manager may consider and consult with the designer. If they feel that the request is justified, the construction manager issues an RFP to the contractor. If not, the construction manager issues a rejection specifying the reason(s) why the request was rejected. The RFC is recorded in the RFC log and in the RFI file. If the contractor later files a claim, then the RFC is included in the claim file as well.

The construction manager should meet with the contractor and discuss the issue whether it is accepted or not.

If the request for change is rejected, then the meeting is to make sure that the construction manager and designer understand the issue and to make sure that the contractor understands why the request was rejected. Let no issue float; they will not just go away. If at some point during the discussion both parties agree that they cannot agree, then contractor is free to file a claim and work the claim through dispute resolution.

Summary of the request for change order:

- RFC is the process through which the contractor can ask for a change that he believes he is due but has not been afforded.

- The contractor's issues should be heard whether the RFC is accepted or rejected.

- The RFC could become a claim if rejected.

Undocumented Field Changes

Purpose: To explain how field changes are ultimately documented.

These do not exist. Any change to the design that the construction manager or his staff knows about shall be documented via RFI, field order, work change directive, and/or ultimately through a change order. Changes to the design must always be approved by the designer of record and may result in a no-cost change order.

Undocumented field changes do not exist!!

Chapter 6

Claims

Contractor Claims

In situations where the contractor feels he has been treated unfairly, the contractor may prepare a claim against the owner for money and/or time. These concerns of unfair treatment may arise from an interpretation of the construction documents, assessment of a change activity as a field order, not a work change directive, or alleged delay to the contractor's schedule by another party, including force-majeure. The contractor has a right of recourse and may request via claim compensation for damages he feels are owed by the owner to make the contractor whole. We will discuss the nature of these claims in detail in this section.

Delay Claims

Owner-Related Delays

Purpose: To provide an understanding of how the owner may cause delay claims and how to prevent these problems from happening.

The construction manager represents the owner. Sometimes the owner is his own worst enemy. Adding major scope changes that take time to design, price, obtain approval for, and implement can adversely affect the project. Another area that becomes an issue that causes delay claims is inattention to document flow and information.

Changes to the design of a project may seem small to the individual requesting the change, such as adding a pump, striping, even to change the location of a door. Oftentimes these changes are rooted in design phase lapses or late owner review. The owner requested something the designer did not pick up on or the request was considered low priority and the requested change was not incorporated into the design. During the review phase, the issue was raised and not addressed or the review just didn't happen and again was missed. During the construction phase, someone looks closely at the design, and there is a crisis because the issue was not addressed during design phase and now, during construction, must be addressed.

These owner-added or -changed design issues are more than twice as expensive to resolve during construction and often result in delaying the contractor because the project now has to wait for design, approvals, and or equipment/material deliveries after approvals are received. The contractor may also lose his ability to be as competitive as he was during the bidding stage.

The other primary owner cause of delays is not attending to documentation in an expeditious manner. At times the construction manager may need to push the owner's staff to complete submittal reviews because the contract prescribed time is nearly over and the submittal needs to be returned to the contractor so the product or equipment in review can be ordered and delivered or corrections to the submittal may be made and resubmitted for approval. Contractors may, over the course of a project, try to claim delay because one or two submittals and/or RFIs have taken longer than specifications allow, but usually most have been handled within the standard of care for time. Keeping the log accurately is the only way to know that information.

As mentioned above, the construction manager represents the owner. That obligation does not mean that the construction manager is to roll over and accept whatever the owner wants to do on a project. The construction manager needs to remember the owner hired him because of his experience and ability to manage his project. He hired the construction manager because he doesn't have the expertise or qualified in-house staff. *The construction manager's obligation is to listen to the owner and then advise the owner of any impacts or risks to the project based on his direction. If the owner chooses to ignore the construction manager's warning or accept the risks that the construction manager has identified, that is his prerogative. At that point, the construction manager goes forward, working on the owner's behalf.*

Summary of owner-related delays:

- Owner-related delays are often rooted in missed design issues or owner-added items during the construction phase.

- Good logs and documents protect the construction manager and the owner from claims stemming from submittal or RFI time overruns.

- The construction manager should advise the owner regarding claims and actions in claim cases, but the owner will decide a course of action, and the construction manager must support the client down the path chosen.

Construction Manager-Related Delays

Another area that becomes an issue that causes delay claims is inattention to document flow. These delays are a result of the various parties, most specifically the construction manager, not being attentive to the length of time that the various parties hold documents in review or that need to be submitted. If an issue comes to the construction manager's attention he needs to follow up and see that the issue is addressed by RFI or submittal in an expeditious manner so there is time allowed for the appropriate party (usually the designer) to respond and return before the issue becomes critical. This holds true for the designer responding in an expeditious time after receiving a document for review and answer.

Usually there is language in the construction document that requires RFIs and submittals to be responded to and returned in a certain amount of time. All parties need to be careful not to suppress questions and/or answers so they delay the construction. These can be a cause for claim submission by the contractor or push a secondary activity onto the critical path of the schedule. Do not sit on RFIs or submittals; move them to their destination as quickly as possible. Make copies for everyone, stamp them, log them in, review them quickly, and then pass them on. Then monitor the time that the designer has them and push to get them back to the contractor within the contract's prescribed time.

The construction manager must take problematic issues firmly in hand, attack them with focus, and then follow up as the issue moves from party to party. Constant and firm pressure is the only way to resolve these types of problems in the fastest possible manner. If the issue is not an option, that is to say it is required for proper operation, the construction manager may obtain permission to allow the contractor, or the construction manager may encourage the owner to pre-purchase any long lead time equipment/material that the added scope requires. This will allow manufacturing time while the issue is resolved.

Regarding claims against the construction manager, keep in mind the following tips:

- Documentation is the construction manager's protection.

- Poor control of documents is most often the cause of such claims.

- Attack problematic issues. Do not let them sit; they will not go away.

Designer-Related Delays

Purpose: To develop within the construction manager staff an understanding of delays to the contract, in this instance those caused by design issues.

No design is perfect. In a later section, we will discuss a principle known as "standard of care." There is, however, a very high expectancy of accuracy. Some issues caused by the designer will affect the schedule, project cost, or both. The three most frequent issues through which the designer causes delays to the contract time are:

1. Excessive time reviewing submittals.
2. Excessive time answering requests for information.
3. Excessive time in producing design upgrades required during the construction phase.

The contract is explicit regarding the amount of time that the designer has to process RFIs and submittals.

The role of the construction manager is to see that:

1. All documents pass through his hands as quickly as possible.
2. To remind the designer before the documents are past their due dates.

If the designer needs more time on one of these types of documents for review, research, and obtaining clarification to the information provided, then the designer needs to notify the construction manager and the construction manager needs to work with the designer to resolve any such issues as quickly as possible. Keep the contractor in the loop as well. That keeps anxiety to a minimum and claim considerations from the front burner.

Design changes increasing the scope of the contract or modifying the work in a major way can also cause delays, but those instances are fairly rare and are usually known about when the redesign or design modification begins. When those types of things happen, bring the contractor into the picture by alerting him or her well ahead of time that the change is coming and that it will be substantial and his response is due as soon as practical.

Be prepared to give a high priority to review of the designer's response to the RFP and to prudently negotiate an agreement and get the recommendation through to approval as quickly as possible.

Designer-related claims

- are usually caused by poor control of documents by the construction manager.

Contractor-Related Delays

Purpose: To develop an understanding of contractor self-imposed problems that causes him/her to fall behind (slip) the schedule.

Very few contractors enter into a contract with the intention of completing the project after the contract-established completion date. Once in a while an impossible date to reach is established by the owner or designer in the document, but that is another problem. Contractors make money if they complete ahead of schedule. That fact escapes many contractors, especially the small, unsophisticated fellows who are not good managers. The most common cause for slippage on a schedule is insufficient manpower, equipment, or materials to press the work forward judiciously.

The most important reason for the resource and cost-loaded schedules is to allow the contractor and the construction manager to track the resources used by the contractor during the course of the project. Man hours used is also known as the man-hour burn rate. If the contract is ten months long and has one hundred thousand man hours needed to complete, then the contractor must have enough manpower to burn ten thousand hours per month. That equates to a crew of fifty-eight men working forty hours per week each month. This is a pretty reliable indicator of production toward the end goal of on-time completion. If the contractor lists three cranes for the project and has only one on site, chances are pretty good that he is slipping or will fall behind before long because there just isn't enough equipment to move the product and equipment into place fast enough to get the project completed on time. If the project is $1,000,000 in value, the dollar burn rate is $100,000 per month. Realize that the first month or two may be low, but for a short duration project like ten months, the curve for spending must be very steep, so if billings persist at $50,000 to $80,000, the contractor is probably in trouble.

Prevention is to be proactive. The construction manager must monitor the schedule on a weekly basis. Review the three-week look-ahead schedule and compare it to the overall project construction schedule, which is where the slippage will first become apparent. Push during the weekly meeting to get back on schedule early. Incorporate strong language in the construction document that allows the construction manager to intervene quickly and ask for a recovery plan and schedule. If need be, call the contractor's superintendent in for a conference about what has been observed. In order not to be adversarial, offer assistance where you can, offer *suggestions* for better productions, and suggest where the contractor may be able to obtain more and/or better people. If contractor has equipment problems, offer *suggestions* regarding rental equipment that could help his equipment shortfall. No matter what the construction manager does in the way of assistance, he is not to direct the contractor or his crews or determine means and/or methods.

Preventing claims of all sorts

- takes diligence in document control,

- takes diligence in project oversight, and

- takes understanding of the budgets, schedules, and perhaps intervention tools.

Owner-Caused Non-delay Claims

Purpose: To provide an understanding of what the owner may do to cause claims and how to prevent these problems from happening.

Owner-related claims come as a result of one action:

The owner refuses to pay for a change to the scope of work.

The construction manager represents the owner. Sometimes the owner is his own worst enemy by adding major scope changes that take time to design, price, or obtain approval for. Often the owner will expect these requests to come at no cost perhaps because "they should have been included" in the design or the contractor's price. The contractor will not include pricing for elements not included in the design. The contractor will bid the lowest-expected quantities. As stated before these owner-added or -changed design issues are often more than twice as expensive to resolve during construction and often result in delaying the contractor because he has to wait for design, approvals or equipment/material deliveries after approvals are received.

Sometimes the owner or designer just doesn't want to pay for the requested change. If the requested added scope is not included in the design or not defined or directly named in the construction design documents, then the owner or the designer have no right to expect a no-cost change and must be advised accordingly by the construction manager. If the owner persists in pressing for no cost, then the construction manager must explain that if he insists, the outcome may lead to a dispute, claim, or litigation and that there is a good chance that he may not be successful should the contractor file a claim for payment. If the owner persists, the construction manager is bound to act in accordance with the owner's direction and prepare for the outcome.

The owner can cause claims

- by not paying for owner directed additions to scope.

- by insisting on acting against the contract, or

- by not being attentive to document flow.

Claims against the Designer

Purpose: See claims against the owner.

Claims against the designer usually originate with the contractor asking a question that either points out what the contractor perceives to be an error in the design or a group of issues that demonstrate some level of incompetence or negligence on the part of the designer. Mistakes in the design are inevitable. That is why a *standard of care* level should be incorporated into the contract with the designer. There are two reasons that an issue causes a claim against the designer. They are as follows:

- The dollar value of the issues that are charged to the designer's errors or omissions (E and O) exceed the sum level of the standard of care. A standard of care is the level of error acceptable in the documents

- There is negligence in the design found on the part of the designer.

Standard of care is the concept that a certain amount of error can be expected and is acceptable in the design. The question is, how much error is acceptable? What is the level of care taken during design? How much care was taken during design review? Was the owner allowed to review the design? The answer to those and other questions yields the care taken to produce a document that meets a standard of excellence, thus the phrase *"standard of care."* That means the designer is allowed a cumulative total of some percent of the total constructed cost in design errors and/or omissions, which causes change orders before the owner begins to assess error and omission related back charges against the designer. Standard of care is a common practice in the architectural design community and is gaining traction in the engineering profession as well.

Errors are those design elements that, even though constructed, will not work. These are constructed prior to discovery. Such an issue would be an elevated slab designed and constructed to a thickness and/or reinforcement not of sufficient strength to support the intended loads or not of sufficient safety factor to meet code. The construction is complete, so the cost of design, demolition delays, and reconstruction to the new design are charged against the designer.

Omissions are design elements left out or omitted from the original design that need to be designed and constructed to make the project meet the original design intent. Because these were not constructed, the cost against the designer will be for the design and will cover any delay costs but not the costs of normal construction. The construction cost would have been in the bid cost had the issue been discovered and incorporated into the bid document.

The cost of the added construction is added value to the owner over the value prior to the work's completion.

Not all errors and/or omissions generate cost and/or time against the designer. Errors and/or omissions that increase the value of the completed project are constructed at the owner's expense.

Only issues that do not generate added value to the owner are true costs against the designer. An issue of error or omission that causes something new to be added that is necessary to the function of the end product that is caught before a negative impact to the project and that costs no more than it would have at bid time will not be charged against the designer's Errors & Omissions (E&O) tab. However, if demolition and restoration must be accomplished at the cost of the owner to add or change an already constructed item, the cost of the demolition and restoration to the point of new construction should be charged, but the cost beyond that point to construct the addition should not be charged.

If constraints or project conditions dictate that an error and/or omission will be constructed at a higher price than that same item would have cost if it was in the document correctly at bid time, the cost differential will be added to the designer's E &O tab. Costs that may be wholly added to the designer's E&O tab usually result from wholesale design changes to the process that cause new construction and modification to existing product to make the already designed process workable. (Existing product in this case would be completed design and/or construction.)

Claims against the Contractor

Purpose: See claims against the owner.

Claims against the contractor are rare. They usually result from purported negligence on the part of the contractor during construction and/or back charges that cannot be resolved at the project site level.

The main reason that claims against the contractor are rare is that virtually all contracts with contractors retain monies (retainage) to incentivize proper completion of the project. The retainage can be used to compensate the owner for uncompleted work, work completed but deemed below standard, monies owed to subcontractors or vendors by the contractor in order to obtain a release of lien statement, and/ or back charges against the contractor. It is usually only after the retainage is exhausted that a claim is filed against the contractor. However, charges that change the contract amount or time are only made via change order even if the change is a reduction in payment or retainage. The other cause of back charges is damage to the owner's, construction manager's or designer's property. Those damages are usually resolved via insurance claims.

A third source of security against the contractor is the contractor's payment and performance bond. Activating the bond usually requires filing suit in court. However, the contractor's bond does provide an additional means of recovering cost when a contractor fails to complete and/or pay for labor or materials.

Claims against the contractor

- are rare because there are three securities that normally protect the owner and prevent claims: (1) retainage, (2) insurance, and (3) payment and/or performance bond; and

- are documented by change order if they result in an adjustment to the dollars or time in the contract.

Delay Claim, Time Extension Requests

Purpose: To provide an understanding of time extensions.

All time extensions are not equal. Some arise from added scope change orders. Some come from force majeure issues, such as weather delays. Some arise from delays caused by late arrival of equipment purchased by the owner. Some are caused by items that arrive late that the contractor purchased. Some arise from unresolved issues that carry on for an extended period of time. Some arise from a third party that slows or stops the contractor's progress on one or more activities. Each issue that arises that could cause an extension of the contract end date needs to be addressed as quickly as possible. As on the cost side, do not let a sleeping dog lie. "The dog that is ignored bites the hardest." Treat each issue individually. When a change order is written, include both cost and time associated with the change.

Remember delay claims are only valid for issues that relate to critical path activities. The most important reason to approach and resolve all issues as quickly as possible is to keep both cost and time under control. An issue that has float available may be pushed onto the critical path if it is not resolved before the entire float is exhausted. At that point, if the issue has not been properly addressed, the owner may be at risk for a delay claim.

Delay claims may be cost associated or not, depending on the issue and how the contract language is crafted. Typically a delay claim could be for time extension only. However, if there is an issue that adds work scope that could be accomplished within the normal contract time and it is not resolved in a timely manner, which forces it onto the critical path, the project may face a cost increase along with a time extension.

Delay claims are handled the same way all other claims are handled, requiring proper explanation and justification.

Time delays and/or time extensions

- must be addressed quickly,

- must be addressed in a change order in the same way as costs are addressed,

- may or may not be cost associated, and

- must be thoroughly justified.

Defending against Claims

Purpose: To present the concept of claims and claims defense.

This entire treatise has been developed to keep the construction manager from having a claim brought against the owner. Now we will explore the process of defense against a claim should one arise. First of all, a claim is an allegation of monies and/or time owed to one party or the other against one of the other parties of the contract. That is to say the contractor, the owner, or the designer feels that one of the others owes him money and/or a time extension because the contract has been in some way modified or breached. Most claims are brought by the contractor against the owner.

As stated in the change order section, claims may be costs and/or contract time changes that have been denied by the owner. It is important to resolve claims at the lowest level as quickly as possible. Claims resolution may be elevated beyond the site and contested at a higher level. Claims are presented to the construction manager so he can check the information to ascertain that all of the following apply:

1. The issue is perceived to not be covered under the contract language or is caused by a perceived breach of the contract in some fashion.
2. The information is complete.
3. The information is presented in sufficient detail to make an intelligent decision.
4. No information is presented beyond that presented at the site level. (If more information is available, it needs to be assessed at the site level first. Changing the method or style of presentation of information is allowed, but the facts may not be modified unless they can be substantiated to have been faulty.)

Claims should only be submitted as a last resort, so rearranging information into a different style that better tells the story should be explored by both parties.

Claims by the contractor usually develop for one of the following reasons:

- Differing interpretations of the project document, between the owners representative, and the contractor. While the construction manager may have knowledge regarding industry-standard interpretation, the construction manager should appeal to the designer to obtain the design originator's definition and intent of the point in question. The construction manager may review both opinions or call for a legal interpretation to settle the issue. If the contractor does not agree with what comes forth from the designer and/or legal review, he may file a claim.

- If a contractor is made responsible for "code" in the construction document and his interpretation of a point in the "code" differs from the designer's interpretation, he or she may take exception and file a claim for protection or restitution.

- Differing opinions between the contractor and the construction manager over values of cost and/or time presented in a request for proposal. The construction manager may appeal to the designer or outside estimator for review of the contractor's proposal. If contractor does not agree with the review, he may submit a claim.

The most important issue in claims on the part of the construction manager is to not take the contractor's presentation of a claim personally. These are business decisions on both sides. All parties need to make the decisions reasonably and in good faith that the other is fulfilling his responsibilities to the best of his abilities.

In a later section, we will discuss the disputes resolution process. Both the contractor and the construction manager need to be prepared to present their cases at every level of the dispute resolution process. It is important to understand that claims are not positive for either side and indicate an inability to resolve issues by both managers at the site level. However, claims happen no matter what our best intentions are.

Compromise is the best way to resolve claims. Chances are this compromise will happen at a higher level anyway.

Claims

- must be justified in writing,

- may result from misunderstanding, disagreement of responsibility, or interpretation differences, and

- may also arise from failure to come to an agreement over time or cost proposal.

The contractor should be bound by contract to continue working while disputes are resolved.

Expectation of Proof

Purpose: To establish the requirement for filing a request for more money and/or time by the contractor.

During the course of almost every project, the contractor asks for more money and/or more time to accomplish the assigned work of the contract. Usually these follow some change such as an answer to an RFI, a changed condition, or a design clarification. The contractor may feel compensation is due, while the construction manager may not. If a situation like this arises, a verification of why the request is being made is required. Often what the contractor feels is a changed condition is actually covered in the project document.

The burden of proof is on the contractor to verify that indeed what he sees as a changed condition is truly outside the scope of work of the contract. It is not unusual for some items, such as a buried pipeline, a communications cable, etc., to be found while excavating. The contract language may state that contractor is responsible to locate all utilities before beginning to excavate, but the contractor may have just started digging. The contractor must follow the letter of the contract regarding notification and utility search and then must prove that the utility in question was not properly located, or was not locatable, and/or was unknown to any usual source.

It is not acceptable for a contractor to come forward and say, "I think I have a problem. You owe me money to fix it." Evidence of the problem must be presented. The burden of proof lies with the contractor.

The burden of proof

- lies with the contractor.

Documentation of Claims

Purpose: To establish how claims are documented and how refutation of claims are documented.

Contractor must establish the basis for a request for monies and/or time. The construction manager or designer may establish this basis by issuing an RFI. The contractor may request additional monies or time by simply writing a letter stating the amount of money and/or time owed and evidence of the fact. If the statement of fact is backed up by evidence that is solid enough, then the construction manager will issue an RFP so a solid basis for payment is established. If the evidence is not substantial, the construction manager may dismiss the evidence as inconsequential, or he may request more information.

The contractor must supply detailed backup for each item listed as basis for payment in response to a request for pricing. The detail is explained in the "Request for Pricing" section of this manual. The construction manager then documents the review of the price submitted for accuracy and fairness. If the response is inadequate or not acceptable because it is deemed unfair for some reason, the construction manager should sit down with the contractor and discuss the issue. This may take several iterations, but eventually the two parties may reach a stalemate wherein neither the contractor nor the construction manager will concede. The other issue that may arise is that the construction manager has rejected the contractor's claim and will not even issue an RFP. In either case, the contractor may file a claim against the owner for monies and/or time he feels he is owed.

At this point, the two parties agree to disagree, and the construction manager simply tells the contractor that if he wishes to go forward, he must submit a claim. The contractor then submits a claim, which includes a claim document with the priced RFP or denial for RFP, along with the necessary (or supplied) backup to the construction manager. The construction manager reviews it to see that what has been submitted is complete and sends it on to the next level as defined in the "Disputes Resolution" section of this manual.

The construction manager then creates a file with the claim number, including copies of all of the documents. At a minimum the file will include meeting agendas and meeting notes from the meetings when the issue was discussed, various proposals from both sides, letters of expectation, and the claim denial.

When a claim is resolved, it requires a written document by the party representing the owner that lays out the terms of the claim resolution agreement, including any monies and/or time allocated in the resolution. The claim resolution document is then turned into a change order and issued as a change to the contract. The construction manager should log the document in as an RFP response in the change order log, list the price of the resolution in the change order amount, add the amount of time (it may be a negative number) in the schedule change column, and fill out the balance of the change order log. Keep a log, and keep the log current at all times!

Documentation of claims:

- lies with the construction manager.

The construction manager documents the process, checks documentation from all parties, and requests more documentation when that supplied is not thorough.

Claim Prevention

Purpose: An executive summary of claim prevention.

This entire document is geared to managing construction projects without having any disputes that cannot be resolved at the site level.

The construction manager is employed by the owner but works as directed by the documents. That is to say the project documents govern the project. The owner has written, reviewed, and approved the project documents. The designer has designed around the project documents, and the contractor has bid to the project documents.

Enforcement of the project documents during the construction phase is the responsibility of the construction manager. Proper management of the project as directed by the project documents is the key to claim prevention. Setting up a proper filing system is essential, and it is essential to file all received originals and copies of all sent documents properly. The project documents may require that the construction manager take a stand contrary to what some deem to be the owner's best interest. In the long run, the documents will serve to protect the owner, so the construction manager must do as the documents say.

The construction manager must be a detail-oriented person when dealing with documents and protocols of notification. Claim cases generally hinge on "who knew what when." The document trail date stamped and forwarded in a timely manner is extremely important.

Managing good progress meetings and getting the contractor's input for the agenda promotes an attitude of teamwork. Holding special meetings to resolve issues as they arise is important as well. If contractor feels that the construction manager is concerned about his (the contractor's) issues, he will be more inclined to go forward in good faith and be more cooperative on issues that need resolution. Getting all of the stakeholders to buy into the project as teammates is important for the morale of the project. When all parties on a project have a good attitude, claims are rarely seen. When one party feels slighted or gets in trouble with either cost or time the whole attitude of the project can be in jeopardy and an issue can be expected, that party can be any one of the stake holders. "One balky horse and the wagon stops rolling forward." The construction manager's time spent in the field and with the owner can be invaluable to sense these issues and bring about quick, early solutions.

Claim prevention

- *is the point of this publication*

- *The construction manager's pre issue efforts can prevent many problems.*

Specific Techniques for Mitigating Weather-Related Claims

Weather is not under control of the designer, the contractor, or the owner. Who is responsible for weather delays? The owner usually is saddled with that burden. How do we protect the owner from costly weather delay claims? It is done by developing strong, fair language in the construction contract document. There are several good strategies commonly used in contract language. Here are a few.

Obtain the local weather data collected over the past several decades, and establish the highs and lows. Use the established high temperatures, low temperatures, high rainfalls, high snowfalls, and excessive wind to establish fair marks for what you define as severe weather that will trigger "severe weather events" for which the contractor can claim damages.

For example, the average low temperature for January is 25 degrees F. The low temperature over the past thirty years is -15 degrees F, and that happened four times. The average low is 12 degrees F. One might use that information to set the contractor compensable limit for abnormally cold weather at 10 degrees F or colder sustained for a period of ten days and 20 degrees F or colder for a sustained period of thirty days. That is to say that if the temperatures from January 2 through January 18 range from 4 degrees F to 9 degrees F and then climb to a balmy 25 degrees F, the contractor may be paid for heating costs for that period of seventeen days. If the temperatures from January 2 through January 29 range from 4 degrees F to 27 degrees F, the contractor is not compensable beyond the initial seventeen days because the threshold of thirty days was not reached. Limits defining a severe snowfall or rain storm can also be defined. Consecutive days of precipitation above a certain level can be defined as well. Tie the limits used to established data. Be realistic and fair when establishing the limits.

A second method is to set a number of weather days as an allowance in the contract. Again obtain the local weather data. Set a realistic threshold to define a limit that may be used when the contractor may choose to close the project. Determine how many days each month have actually been weather days over the time of the data. Set the number of allowance weather days that the contractor must allow for in his bid. Be sure and check that the master construction schedule submitted by the contractor includes those days.

For example, say over the past thirty years, March has averaged nine days of rain or snow that yielded beyond a quarter inch of precipitation. April has averaged six days, May four days, And June zero days. You may put in the contract, "The contractor shall not be entitled to additional weather days unless the number of rain/snow days where the amount of moisture exceeds a quarter inch is: March—ten days; April—eight days; May—five days; June—one day," and so on for each month of the year.

A third is to allow days to be added for weather but no cost. The contractor may be granted a time extension but will not be compensated for costs incurred due to inclement weather. This strategy will probably work for moderate inclement weather but will probably not work if severe conditions occur or if even moderately poor conditions occur over a sustained period of time.

Chapter 7

Dispute Resolution

Dispute Resolution Process

Purpose: To clarify points of difference and establish ground rules for dispute resolution, which establish lines of communication for discussing points of dispute and hopefully find a common ground on which to resolve each dispute.

Dispute resolution has been brushed upon in several sections but bears repeating. As previously stated, The first rule of dispute resolution is: *to resolve all disputes at the lowest possible level and as quickly as possible.*

Dispute resolution meetings are held at several levels beginning at the project level and ending at the highest level of the contract's or owner's choosing. Most disputes will be solved in the field at the field office. All formal dispute resolution meetings will be governed by agenda, which will be developed by either party involved in the dispute. Both sides have opportunity to include items for discussion. Items of dispute should be discussed openly and respectfully; hostility has no place in a professional setting. Oftentimes agreeing to disagree is the only solution at a given level in the dispute resolution process.

The dispute resolution process begins at the project level. A dispute may arise between an inspector and a foreman or an inspector and a field superintendent. The meeting at this level may occur in the field at the very point of the dispute. The discussion at this level should be civil and fact based on contract requirements. If that level cannot resolve the dispute, they refer the issue to their immediate superiors and allow that level an opportunity to resolve the issue. The site engineer for the contractor should meet with the construction manager's team engineer and discuss the issue, using the contract document as the foundation of the discussion.

If the people on site cannot resolve the issue, it may become a claim, and the contractor must provide written justification from the specifications supporting his point of view regarding the claim to the construction manager. The contractor must also provide a detailed breakdown of all cost- and time-related issues of his claim. The project-level owner's representative, usually the construction manager, should review the contractor's claim presentation documents and request any information that is necessary to evaluate the claim properly. Meetings may be held to allow the contractor to present his justification, cost, and schedule information. The construction manager should ask questions and be certain that he understands the claim and all of the information presented. Another meeting may be held to discuss possible resolutions, including acceptance, possible compromise, and/or rejection. Several meetings may be held to discuss compromises. If none can be reached, then the claim is considered irresolvable at the project level and is handed to the next level up the dispute-resolution process. Detail should be sufficient at this time to allow the next level in the process to understand the case and meet to resolve the issues.

Meetings at management levels may focus on many issues but should generally remain true to the principles of the contract document. If the top levels of management cannot reach a resolution, then the dispute will be elevated to a dispute-resolution party. This may be a disputes resolution board (DRB), arbitration, mediation, or court as the project documents require.

The purpose of the various levels of the dispute-resolution process is to prevent issues elevating beyond the levels of the parties involved. Disputes that go into the court system are wasteful of time and money for all parties involved.

There are several methods used to resolve issues short of adjudication. Among them are the following:

- A dispute resolution board is installed by all parties at the beginning of the project with representatives of all parties on the board. There are usually three to five members on the board, with one or two chosen by each party, and those choose a third or fifth who is acceptable to both as the neutral member. This group usually hears all disputes elevated beyond the parties and is empowered to reach a binding decision on all issues that they hear.

- Arbitration, either binding or nonbinding, is a dispute-resolution process whereby a disinterested third party, often an expert, appraises all sides and assesses liability on each party. Binding arbitration requires that the parties must abide by the assessment. Nonbinding arbitration allows the parties to accept or reject the assessment.

- Mediation is a dispute resolution process whereby a disinterested third party mediates the parties to a mutual compromised settlement. Mediation provides structure, a timetable, and a dynamic (the disinterested party) that normal negotiations lack.

- Litigation is for most stakeholders the last resort and least desirable.

The actual dispute resolution steps may be as follows:

Step Contractors' Representative	Owners' Representative
Field foreman or field superintendent	CM inspector
Site field engineer	CM resident engineer
Project superintendent	CM construction manager
Project manager	PM manager of construction
Vice president	Project manager
President	Owner's manager

7. Dispute Resolution Board, Arbiter, or Mediator, Court. (See "Dispute Resolution Meeting.")

Chapter 8

Testing, Startup, and Commissioning

Lock-Out/ Tag-Out Procedure

Purpose: To establish a safe procedure for any and all trades to access any system or piece of equipment for work after a system or piece of equipment has been deemed operational.

Unless the owner requires that the designer or construction manager develop a lock-out/tag-out procedure prior to energizing the first system or piece of equipment, the contractor will present a lock-out/tag-out procedure that allows any and all crafts that may need access to the working side of a system or piece of equipment to access it safely, making sure that system or piece of equipment cannot be made operational while a craftsman is working on it.

The procedure must require communication of the status of all locked-out pieces of equipment during the contractor's or owner's off-work hours on which any worker has his safety device in place.

The procedure must prevent even accidental energizing or pressuring of a "locked-out" system or piece of equipment.

The procedure must allow for energizing or pressurizing a system or piece of equipment that is shown status-wise as ready to perform but remains in a locked-out situation during off-work hours of the party that isolated the system or piece of equipment. That is, the procedure must allow for overriding in extreme emergency and explain in detail how the override will be accomplished safely.

Pre-Operational/ Functional Testing

Functional Testing

This phase of the project marks the beginning of the end of the construction phase of the project. It is the time when there is more left to test than there is to construct. The contractor is responsible to provide a working product. He knows how to do that. It is simply to activate each piece installed to see if it works and will not fail under a real, live load achieved at some level above normal working conditions. Piping is usually tested at 50 percent above normal working pressure or as specifications direct. Electrical systems are tested at capacity.

Hydraulic/Pneumatic Pressure Testing

Purpose: To provide guidance for testing of piping and instrumentation systems.

The contract documents will provide specific testing pressures and test durations for the various piping systems. However, there is very little, if any, guidance regarding how to document the testing. Generally the project documents require that the contractor do the following for system checking and testing:

1. Define the various systems within the plant design. Each of these systems is relatively isolated and functions separately from the other systems.
2. Develop a book with tabbed section for each operational system.
3. Good results can be achieved with a three-ring binder tabbed for each system. For large plants, use a three-ring binder for each system with tabs for each type of device, such as instruments, valves, equipment, etc.
4. Determine, from the project documents, everything that belongs with each system. Most of the mechanical components can be identified on the Piping and Instrument Drawings (P and IDs) and the electrical from the single line schematic drawings.
5. Assemble a table for each system that identifies each item that requires testing.
6. Create a separate test sheet or sheets for each item to be tested, be it a piece of equipment, a section of pipe, or a system of flow.
7. Write a test procedure for each item requiring a test. The specifications will give the parameters for this.
8. Mark each sheet or set of sheets according to the table assembled in step three.
9. Determine the test requirements for each item in the table.
10. Write the requirements on each test sheet in the appropriate location.
11. The contractor must notify the inspector twenty-four hours prior to a system walk down and test because *each system walk down and test must be witnessed and requires two signatures, the contractor and the owner's representative for acceptability. Acceptance signatures and dates will be affixed when all items are determined to be in place per the project documents.* (This requirement protects both the contractor and

the owner and helps ensure that all systems are indeed complete and functioning properly.)

12. Walk each item down, marking off each item required, checking location and orientation as applicable.

13. The test plan will be implemented and the equipment, piping, and/or system will be tested and test results recorded.

14. System problems that require rework and/or test failures will be noted in the book and the results of the subsequent inspections and/or test noted and dated. ***Acceptance signatures and dates will be affixed when all items are determined to meet the test criteria of the project documents.***

15. Two sets of books minimum should be made and signed by both the contractor and the owner's representative witness. One set is the owner's and the other belongs to the contractor. The contractor's set will be turned over to the owner at the end of the project with the as-built "red line" drawings.

16. Operational testing will be accompanied by a day-to-day log of activities that happened in regard to each piece of equipment or system. Each item will have a place for the log entries, and a general log will be posted in the back of each book. (A copy of the log may be kept in the contractor's book.)

17. On the last page of the book, a signature page will be developed where the contractor and the owner's representative attest to the entries in the books.

Mechanical, Electrical, Leak, Flow, and System Testing

All of the testing will be managed according to the procedure described in the "Functional Testing" section found on the previous page.

Test Reports

Purpose: Checking the quality of the installation of the equipment, instrumentation, electrical, and piping.

The project construction schedule will have an item called "startup and testing" or "testing and commissioning." This phase of the project must have a schedule of activities of its very own. One schedule activity, whatever it is called, is simply not enough. The contractor is responsible to provide a working product in which the individual pieces perform properly. The function of the testing phase is to test the individual components of the system to know, before commissioning, that the system as a whole is capable of doing what it was designed to do.

Each of the following components is to be subjected to a test as prescribed by the specifications, ASTM, ASME, AWWA, API, ANSI, manufacturer's standards, or some other standard to ensure that the component was installed correctly and will function safely within the daily working limits.

1. Individual pipe systems, including pipe, valves, seals, gaskets, fittings, devices, and attached instrumentation should be tested to pressure and duration.
2. Instrumentation—all instruments will be bench calibrated prior to installation with certificates of recent calibration.
3. Electrical motors should be bumped to determine rotation direction after wiring and before equipment is attached. Then when they are coupled to the equipment, they will be tested under load for operational amperage draw.
4. Electrical gear should be tested for functionality and current flow under load conditions.
5. Equipment such as pumps that are coupled to drive motors should be aligned to the motor prior to attaching piping and then after flanging in piping to verify that

the piping is not affecting the equipment with undue horizontal or vertical stress.

6. Equipment such as pumps, filters, heat exchangers, chillers, presses, etc., will be tested for function at low and high flows through their standard range of operation. Heating, noise, vibration, and how they affect the operation of other equipment will all be monitored and documented.

7. Program logic will be tested for functionality, safety, and reliability.

8. Complete piping systems will be flushed using conical screens at each pump intake to protect impellers and/or vanes from damage.

9. Safety devices such as over pressure alarms and shutdown devices will be field tested to operating point. Misalignment detectors and high/low sensors will be tested to make sure that they work at their proper settings.

Samples of test report sheets are to be filled out for each test and kept in the system binder. Binders should be kept for each operating system (see step three of "Hydraulic/Pneumatic Pressure Testing"). Two sets of test records should be kept. Set one is for permanent plant operations records and field use. Set two is for the permanent record and kept in an archive of project information. (This set could be the set the contractor keeps during testing, which is given to the owner with the redline drawings.)

All tests are to be planned and discussed during the daily startup and test meeting held each morning at the beginning of day shift. All tests are to be witnessed and signed by an owner's representative.

All automatic starts and stops are to be tested under running conditions to ensure the safety and security of each system before commissioning begins.

During commissioning, each system will receive a full shakedown to ensure that it will function as designed without mishap, leaks, or mechanical or electrical failure.

Commissioning conditions are to be recorded in the commissioning book that has been set up for each system. The commissioning book is similar to the startup book.

The contractor will transmit the testing and commissioning book for each system to the construction manager. The resident engineer will check for completeness, accuracy, and acceptability of each system before final closeout of the project.

System Startup and Integration

(primarily for plant operations)

Startup of tested systems is a business unto itself. Any one of several parties might be best suited to start up any given system. However, the more complex and edge of technology the system is, the tighter the operating parameters, or the more complex the system integration is, the more important it is that the designer play a major role in the startup, even to including the startup activity as a portion of the designer's contract. Choose a responsible party for startup when the contractor does not know the operating parameters and how each piece of equipment is configured to affect the final and/or optimal operating capacity. The designer knows how the plant is to work as an integrated whole. This is especially true of plant facilities. In all cases, the owner should have a large role in the startup of a plant, roadway system, or any other facility.

Upon the completion of the pre-operational or functional testing, startup of systems begins. The system must be functional at some level to test the various components, including equipment and the system instrumentation and controls. The startup process serves to integrate all of the various components of each system into a cohesive operating unit. When all of the systems have been integrated, they are in turn integrated into the functioning process they were designed to be. When the process is started up, the startup team balances each system in the process to function at the desired or optimum rate.

Once the process is started up and balanced to operate at its desired performance level, a series of test runs are performed to ensure that each system will go off line and come online and seek the established performance under stop-and-start conditions. When these tests are completed, the startup crew, along with the embedded operation team, will begin the commissioning of the integrated process system.

Commissioning

(Mechanical, Electrical, Instrumentation and Controls, Flow Rates, Hydraulic, System Functionality, System Operation)

Purpose: To define the roles of the various participants in the interim stage between construction and operations.

The final project responsibility for the construction team is to accomplish a functionality test of the system as designed and constructed. Once this is completed, the system is taken jointly by the designer and contractor through a startup phase that provides operational checks of all of the systematic components, such as level indicators, level alarms, automatic shutoffs, and safety devices. Each system is then started manually to check function interface and automatically to ensure that sequencing and timing of components coming online is proper.

After each system is checked operationally, then the entire plant is integrated into a startup regimen to ensure proper integration and programming of the plant operation as a whole. At this point the plant should be started and run through all range of operation under all available configurations to yield a range of data for all operating conditions. The start-up and owner team will decide which configuration will be optimal for best plant operation will be best for full operation. The start-up and owner team will then start the plant and run it under operating conditions for a period of time, the time period is usually defined in the construction document. If the plant performs as expected then the plant has been commissioned and may be turned over to the owner.

The contractor will assist in that some of the contractor's people may be employed to assist as directed by the designer's startup team. Once this operational integration and programming check is complete, the operational test period begins and will continue for a set period of time determined by the project documents. Per the section titled "Functional Testing", a daily log will be kept with the pertinent daily activities and what happened during any testing, such as items tested, problems found, and decisions and corrections made.

Chapter 9

Project Completion and Closeout

Substantial Completion

Substantial completion is commonly defined as "the point at which the project is able to be utilized for its intended purpose."

Another important consideration is that contractor will not substantially interrupt operation of the completed portions to achieve final completion status of the project. At this point, time ceases to be charged against the project clock, and liquidated damages cease to be charged and/or cease to be looked to as a remedy for the owner to recoup lost revenue. Federal, state, and/or local law or regulation may define substantial completion and how it is to be treated and handled, be sure to know and understand the rules before you begin closeout procedures for the project.

When contractor feels that he has completed the project, he should file a request for substantial completion. The construction manager will determine if the project has reached a point where it meets the definition of substantial completion. If the project cannot be used for its intended purpose, the construction manager will deny the request. If the project is ready, the construction manager will set up a date for a final walkthrough. The following people should attend or be represented at the final walkthrough:

- The owner

- The contractor

- The designer

- The construction manager

- The lead inspector

During this final walkthrough, a list of items to complete known as the punch list is developed. The punch list (see "Punch List Development,") is the last list of work to be completed that the contractor will receive. The construction manager needs to make sure everybody understands this and is thorough in their review of the project.

When the final walkthrough is complete, the construction manager will take the list of work items and create a formal punch list that will be attached to the formal notice of substantial completion and given to the contractor.

Make sure

- that the project meets the project document's standard for substantial completion and

- all parties are in attendance at the final walkthrough.

Punch List Development

Purpose: To develop a consistent definition of "punch list" and how punch lists are assembled.

Punch lists are the final list of items that need to be addressed prior to the contractor having his completion ticket punched. Once the final punch list walkthrough is completed and the punch list is developed, all parties agree that there will be no more for the contractor to do. That is to say when the punch list is complete, the contractor has completed the work.

If the list has work items on it, then it is a "work tasks to be completed" list not a punch list. "Rubbing a concrete wall" is not a punch list item. "Touch up the rub surface on a wall" is a punch list item. "Backfill valve box" is not a punch list item. "Place correct lid on valve box" is a punch list item.

Punch lists are not meant to be "work to be completed" lists at the end of the contract time. A preliminary punch list should be walked before testing starts to make certain that the project is ready to energize the process system as a courtesy to the contractor, but the final punch list may well be made during startup or after the contractor has completed the functional preoperational testing. The contractor should have already walked a punch list of his own at least twice before the construction manager is contacted for a final punch list walkthrough.

The final punch list walkthrough should be made by the contractor's representative, the construction manager's representative, the designer's representative, the owner's representative, and if possible, an operation's representative. The punch list walkthrough may generate several punch lists, one for each system or perhaps one for each structure.

Punch List Item Check-Off Completion

Once the final punch list walkthrough is completed, all parties agree that there will be no more lists generated for contractor to do. The list(s) are typed on an established form, and all the parties that walked down the project will sign the list. The list is given to the contractor and the inspector or designer's representative, who will check the items as completed. The contractor will address each item on the punch list and each day inform the inspecting party what items are ready to be inspected and checked off of the list. The contractor's representative and the inspecting party will observe and initial in the appropriate box that the item was properly addressed. When each and every item has been properly addressed, the punch list is complete. The original and a copy go into the system test report notebooks. The contractor may retain a copy if desired.

Project Completion
Release of Liens

Purpose: To demonstrate the proper documentation for protection of the owner as a project is completed before it is finally accepted.

The construction manager reviews the document to see what is required for project closeout. The construction manager may require the contractor to obtain lien releases from all subcontractors and vendors under contract to him or her during the course of the project and submit them with the request for final payment. Some contracts will require that releases of lien are with the periodic pay applications as work is completed. Final payment will not be made until all lien releases are delivered to the construction manager and the construction manager has reviewed them and is certain that there are none outstanding. The contractor may be required to close out all permits and submit proof of having done that with request for final payment. All warrantees must be given or as required assigned to the owner.

Law may protect the owner from liens on owned property. The owner should have no encumbrances on owner properties, works, or improvements as a result of the project.

Final As-Built Schedule

The contractor should be required to submit a final record schedule reflecting all changes and actual time frames in which various elements of the project were constructed. The as-built schedule will be submitted with or prior to submittal of the final invoice. Final payment will not be made until said schedule is accepted by the construction manager.

Final Record Drawings

The contractor should be required to submit a set of final as-built "red line" drawings reflecting all changes and variances from the original design for all elements of the project. The as-built drawing set will be submitted with or prior to the final invoice. Final payment will not be made until said as-built drawings are accepted by the construction manager. All final test results with witness signatures will also be turned over to the construction manager along with the red-line drawings

Warranties

Warranties required in the construction documents are to be assigned to the owner and delivered to the construction manager for inclusion with the final project document package that the construction manager is preparing for the owner. These warranties include warrantees for equipment, roofing, coatings, materials, etc., required by the contract. All warrantees will be properly assigned to the owner.

Request for Final Inspection

The contractor will submit a request for final inspection prior to submittal of the request for final payment. This inspection will consist of the contractor, construction manager, designer, and owner's representative. This inspection may be concurrent with the punch list inspection. Final payment will not be entertained until the contractor has made the request for this inspection. The final invoice will not be entertained until the following has happened:

- All record documents have been delivered to the construction manager.

- All warrantees have been assigned and delivered to the construction manager.

- All liens are satisfied and proof of such has been delivered to the construction manager.

- All operation and maintenance manuals are submitted and approved.

- All spare parts required by the construction contract document have been delivered to the construction manager.

- All items on the punch list are completed and signed off.

Notification of Completion and Request for Release of Retention

The contractor should submit a letter of notification of final completion and request the release of any owner-retained funds. The contractor will at this time return the punch list marked as completed by the construction manager's inspector. The contractor will also deliver to the construction manager all required lien releases from each subcontractor and vendor in his service during the course of the project and proof of all permit closeout required by the document. Final completion will not be considered until the completed and checked punch list, and the lien releases, redline drawings, O&M manuals, and the as-built schedule have been received by the construction manager. At the time of acceptance of the final completion notification, the construction manager will approve the project and recommend that all owner-held retainage be released. Some entities may require advertisement and a time period prior to final closeout and release of retainage.

In Conclusion

The intent and purpose of this book is to provide for the use of others, methods that I have learned in constructing, preventing claims, and completing successful construction projects.

- on time,

- in budget,

- well constructed,

- with no claims.

In order to accomplish these goals the construction manager must know the construction documents, have a good idea of construction industry standards of practice, and like working with other people to accomplish a common goal. These methods along with spending time in the field walking around, observing and discussing issues will keep projects going forward with less chance of irresolvable problems arising.

These are four objectives that all of us in the industry strive for. I have had experience modification rates (EMRs) below .3, with incident rates to match. I have managed $100 million projects with zero claims and change order values under 1 percent. I have seen a bid bonanza for a client at savings under the engineer's estimate of over 20 percent. The methods work.

The information in this book will help you be successful at starting, managing, and completing construction projects.

All you have to do is work hard (eat lunch on your day off) at your four tasks, understand what you are doing and walk the job regularly!